Quantitative Applications in the Social Sciences

A SAGE PUBLICATIONS SERIES

Quantitative Applications in the Social Sciences

A SAGE PUBLICATIONS SERIES

Series/Number 07-164

ASSOCIATION MODELS

Raymond Sin-Kwok Wong

University of California, Santa Barbara
Hong Kong University of Science & Technology

Los Angeles | London | New Delhi
Singapore | Washington DC

For information:

SAGE Publications, Inc.
2455 Teller Road
Thousand Oaks,
  California 91320
E-mail: order@sagepub.com

SAGE Publications India Pvt. Ltd.
B 1/I 1 Mohan Cooperative
  Industrial Area
Mathura Road, New Delhi 110 044
India

SAGE Publications Ltd.
1 Oliver's Yard
55 City Road, London,
  EC1Y 1SP
United Kingdom

SAGE Publications Asia-Pacific Pte. Ltd.
33 Pekin Street #02-01
Far East Square
Singapore 048763

Printed in the United States of America

Library of Congress Cataloging-in-Publication Data

Wong, Raymond Sin-Kwok.
Association models/Raymond S. Wong.
    p. cm.
(Quantitative applications in the social sciences ; 164)
Includes bibliographical references and index.
ISBN 978-1-4129-6887-4 (pbk.)
    1. Social sciences—Classification—Statistical methods. 2. Social sciences—Statistics. I. Title.
H61.2.W66 2010
300.72′7—dc22   2009037026

This book is printed on acid-free paper.

10   11   12   13   14   10   9   8   7   6   5   4   3   2   1

Acquisitions Editor:	Vicki Knight
Associate Editor:	Lauren Habib
Editorial Assistant:	Ashley Dodd
Production Editor:	Brittany Bauhaus
Copy Editor:	QuADS PrePress (P) Ltd.
Typesetter:	C&M Digitals (P) Ltd.
Proofreader:	Dennis W. Webb
Indexer:	Rick Hurd
Cover Designer:	Candice Harman
Marketing Manager:	Stephanie Adams

CONTENTS

ABOUT THE AUTHOR

Raymond Sin-Kwok Wong is Professor of Sociology at the University of California, Santa Barbara, and Head and Professor in the Division of Social Science at the Hong Kong University of Science and Technology. His areas of research interests include inequality and social stratification, sociology of education, comparative research, quantitative methodology, and economic sociology. His most recent book, *Chinese Entrepreneurship in a Global Era,* is an edited volume published in 2008.

SERIES EDITOR'S INTRODUCTION

Social and public opinion surveys typically ask questions with responses recorded categorically. These response categories can be purely discrete or ordinal. Let us consider a classic data table, the Midtown Manhattan data of mental health and parents' socioeconomic status (SES), analyzed by Leo Goodman, Clifford C. Clogg, and many others.

Parents' SES	Mental Health Status			
	Well	Mild Symptom	Moderate Symptom	Impaired
A (high)	64	94	58	46
	(48.5)	(95.0)	(57.1)	(61.4)
B	57	94	54	40
	(45.3)	(88.8)	(53.4)	(57.4)
C	57	105	65	60
	(53.1)	(104.1)	(62.6)	(67.3)
D	72	141	77	94
	(71.0)	(139.3)	(83.7)	(90.0)
E	36	97	54	78
	(49.0)	(96.1)	(57.8)	(62.1)
F (low)	21	71	54	71
	(40.1)	(78.7)	(47.3)	(50.9)

A common way for social scientists analyzing data such as these, regardless of their disciplinary origins, is to see whether the two factors (or more than two factors in multiway tables), mental health status and parents' SES in the current case, are related. Or, in statistical parlance, whether an independence null hypothesis can be rejected. Any basic statistics course will have taught the use of the Pearson chi-square test and of the likelihood ratio test. Simply eyeballing the table of frequencies (those in parentheses are expected frequencies in the table above) would not cut it. If we use F_{ij} to indicate the expected value of the observed frequency f_{ij} in row i and column j in the table, the F_{ij} under the model of independence (no association between parents' SES and mental status) is then given by

$$F_{ij} = \frac{f_{i+}f_{+j}}{f_{++}},$$

where f_{i+} gives the column total for the ith row, f_{+j} gives the row total for the jth column, and f_{++} gives the grand total of the entire table. To test the hypothesis of independence, we compute the Pearson χ^2 statistic and the likelihood ratio statistic L^2:

$$\chi^2 = \sum_i \sum_j \frac{(f_{ij} - F_{ij})^2}{F_{ij}} \quad \text{and} \quad L^2 = 2 \sum_i \sum_j \ln\left(\frac{f_{ij}}{F_{ij}}\right).$$

Applying these formulae, we obtain a Pearson χ^2 statistic of 45.985 and an L^2 of 47.418. With the degrees of freedom of 15 (= the number of rows minus 1 times the number of columns minus 1), we reject the null hypothesis of independence at any conventional significance level and conclude that mental health status and parents' SES are not independent of each other; or put differently, they are somehow *associated*.

However, although we know that they are somehow associated, we have not fully used all the available information to explore the form of their *association*. Association models, which are a type of log-linear models, are exactly set up for this purpose. The tests performed above are tantamount to estimating a main effects only log-linear model. Using a uniform association (also known as linear-by-linear association) model,

$$\ln F_{ij} = \lambda + \lambda_i^A + \lambda_j^B + \beta U_i V_j,$$

where the first three terms give a main effects log-linear model with the additional term capturing the association between the two observed sets of scores (i.e., 1–4 and 1–6) of the two variables, we obtain a Pearson χ^2 statistic of 9.732 and an L^2 of 9.895 on 14 df, thereby retaining the null hypothesis of the existence of linear-by-linear association, merely using just one more degree of freedom for the β parameter. By now, the reader must have been impressed by the power of association models.

As one of the key researchers who have not only applied association models in own research but also contributed to the literature on the method of association models, Raymond Wong has written a book that is indisputably sorely missed by the series. He takes us on a journey from where we left off above to many more types of association models such as row effects models, column effects models, row and column effects models, multiplicative row and column effects models, and many other varieties, including models for multiway tables involving multiple factors.

In the example above, we included a term to capture the association between two sets of arbitrarily assigned scores, but the values of which do not have to be fixed as such, and can be estimated by the model. To learn this and other exciting and useful types of association models for our work, simply embark on the journey with Wong as the guide, through the wonderland of association models in a book that is a uniquely and squarely indispensable addition to our series.

—*Tim F. Liao*
Series Editor

PREFACE

Like many works in a well-developed field, this monograph has a long gestation period. Facing the already voluminous literature on association models, I initially saw little ground for new contribution. As I learned and discovered more about the use of such models from my own research, however, I gradually became aware of the need for a coherent development, integration, and synthesis of this particular family of statistical models in the analysis of cross-classification tables. Standard statistical textbooks generally provide little discussion about association models. Even if they do, the expositions are rather limited because they tend to focus exclusively on the analyses for two-way rather than multidimensional cross-classification tables, which are more realistic for most social science applications. The lack of a systematic exposition of the practical use of these powerful and parsimonious models also became apparent to me in the classroom. I was frustrated by the repeated complaints of students who took my graduate statistics course on categorical data analysis about how difficult the material on association models was compared with other sophisticated statistical models covered in the class. Thanks to their honest feedback, the idea for this volume was born.

Acknowledgments

I am grateful to my graduate students in the University of California–Santa Barbara and the Hong Kong University of Science and Technology (HKUST) over the past decade for their patience and comments on various topics covered in the present monograph. I am also grateful to Tim Liao, the QASS series editor, for his encouragement and support. Thanks are due, too, to the two anonymous reviewers for their critical but enabling comments and suggestions. Last but definitely not least, I thank Richard Poon, my teaching assistant at HKUST, who not only pointed out some of the mistakes in the earlier drafts but also introduced the *gnm* module in R to me. This helps resolve my long-standing dissatisfaction with the common practice in the field of not reporting the (asymptotic) standard errors of estimated association parameters. Finally, I dedicate this book to my wife,

Ming-yan, and my son, Hanwey, for their support, encouragement, and understanding.

The author and SAGE gratefully acknowledge the contributions of the following reviewers:

Yu Xie, *University of Michigan*

Kazuo Yamaguchi, *University of Chicago*

List of Tables

List of Figures

CHAPTER 1. INTRODUCTION

Many social science data are naturally organized in cross-classified tabular formats. For example, in sociology, gender and/or racial differences in the relationship between education and occupation, friendship patterns within social networks, cross-national and/or temporal changes in assortative mating; in geography, changes in the characteristics of urban neighborhoods over time and temporal variations in interstate or interregional migration flows; in economics, temporal trends in import/export trade in the global economic system; in political science, changes in the relationship between class positions, party identification, and voting over time; and finally, in psychology, experimental data on stimulus recognition and stimulus generalization. Although our substantive concerns in exploring their systematic relationship may appear straightforward, the application of appropriate statistical tools to decipher and interpret the meaning and complexities of the relationships involved can sometimes be difficult, especially for beginning researchers.

Various attempts have been made in the past to capture association between row and column variables in tabular formats. For example, if one makes an assumption that the variables involved are ordinal in nature, a number of ordinal measures of association, such as Cramer's V, Lambda, Goodman-Kruskal tau, Kendall's tau-b, Kendall's tau-c, gamma, Somer's D, Pearson's r, Spearsman's r, uncertainty (entropy) measure, and eta (η), have been proposed. However, not only do none of the above measures of association offer the natural transformation of the odds ratios (or their logarithms), they all undesirably incorporate information from the marginal distributions as well (Clogg & Shihadeh, 1994, p. 19). Thus, tables with identical odds ratios would yield different measures of association due to differences in marginal distributions (see Agresti, 2002; Bishop, Fienberg, & Holland, 1975; Fienberg, 1980; Rudas, 1997; Chapter 2, this volume, for details about the calculation and properties of the odds ratios). Perhaps more important, these single measures of association often do not provide an adequate description of the extent of association in the cross-classified table(s), especially when the number of row and/or column categories becomes large.

Instead of using the above descriptive measures, an alternative strategy is the development of *statistical* measures of association that are empirically derived and therefore can be formally tested. While the development of log-linear models (Bishop et al., 1975; Fienberg, 1980; Haberman, 1978) provides an important way to understand the relationship among several categorical/ordinal variables, the task of interpreting those *unstructured* interaction parameters can be demanding in multiway cross-classified tables and/or when the number of categories for each variable increases, yielding far too many parameters to decipher (see Goodman, 2007, for a nontechnical but insightful introduction of the usefulness of log-linear models in categorical data analysis). Through the pioneering works of Leo A. Goodman, the late Clifford C. Clogg, and Otis Dudley Duncan, and their collaborators, we now have a large repertoire of statistical models, *association models* in particular, that are well suited for such analyses. While association models have been widely used by stratification researchers, especially in the study of social mobility and assortative mating (to cite a few, Breen, 2004; Grusky & Hauser, 1984; Hout, 1988; Smits, Ultee, & Lammers, 1998, 2000; Wong, 1990, 1992, 2003b; Xie, 1992; Yamaguchi, 1987), such techniques have yet to become widely diffused into other social science disciplines. This is partly because most of the important statistical and formal contributions are scattered in technical journals, and their expositions tend to concentrate exclusively on simple two-way cross-classification tables. With the exception of Clogg and Shihadeh (1994) and Wong (2001), there has not been any systematic effort to integrate the *family* of association models into a single, coherent framework.

The present monograph, *Association Models*, is an attempt to fill this important gap. Through a careful exposition of a class of association models that examines the *underlying* structure of odds ratios, this book offers a comprehensive and unified framework to analyze and understand any social or natural science data that are organized in cross-classified tabular formats. Readers who have general knowledge of regression models and/or generalized linear models should have no difficulty in understanding the materials covered. Knowledge about log-linear models is preferred but not required (Agresti, 2002; Fienberg, 1980; Powers & Xie, 2000). The monograph can be viewed as a natural extension of a number of other monographs already in the *QASS* series, namely, David Knoke and Peter Burke's (1980) *Log-Linear Models* (#20), Michael Hout's (1983) *Mobility Tables* (#31), Masako Ishii-Kuntz's (1994) *Ordinal Log-Linear Models* (#97), and Tamas Rudas's (1997) *Odds Ratios in the Analysis of Contingency Tables* (#119). Some readers may find it beneficial to read them alongside as well.

The book is organized in the following manner. Chapter 2 offers a review of a series of association models that are particularly useful in analyzing two-way tables. To aid our interpretations, the chapter also provides a strong emphasis on the *structure* of odds ratios underlying different specifications of association models as well as their systematic relationships with each other. The chapter ends with two illustrative examples, drawing on public data tabulated from the *General Social Survey* (*GSS*) from various years (Davis, Smith, & Marsden, 2007). After a thorough discussion on how to analyze two-way tables that forms the foundation for analyzing complex tables, the remainder of the monograph deals with the analysis of multidimensional or multiway tables. In particular, statistical models that are generally classified as *multidimensional association* models for multiway tables (Wong, 2001) will prove to be extremely useful and flexible in analyzing complex interaction patterns in multidimensional cross-classifications.

Chapter 3 examines conditional independence and partial association models in higher-order multidimensional tables. In this situation, there is no need to model three-way or higher-order interaction among variables. The chapter first examines ways to decompose various two-way interaction parameters into simpler but substantively interpretable components and ends with two illustrative examples that use the models introduced to gain additional insights about the underlying association patterns.

Chapter 4 examines situations when both two- and three-way interaction parameters cannot be ignored. In addition to the introduction of various statistically powerful layer effects models, it introduces a similar class of multidimensional association models to understand the complex patterns of association. Two examples, one involving grouping variables (gender and time) and the other involving temporal variations over time, will also be presented for illustrative purposes.

Chapter 5 provides two additional examples to demonstrate further the usefulness of association models in social science applications. The first one deals with a common problem encountered by all social science researchers: the question whether certain categories in a table (rows and/or columns) can be combined (Gilula, 1986; Goodman, 1981c). While the problem may seem trivial or minor at first sight, it can provide additional insights about the flip side of an issue that has often been ignored by empirical researchers: the introduction of distortions and biases derived from *inappropriate* aggregation of row and/or column categories (for further details, see Wong, 2003b). The second illustrative example introduces another possible use of association models: as an optimal scaling tool, that is, how to obtain scaling of categories from their underlying association patterns in multidimensional tables (Clogg & Shihadeh, 1994; Smith & Garnier, 1987). Both examples use, extend, and expand previous literature on the same subject matter.

The concluding chapter (Chapter 6) discusses how some recent developments in categorical data analysis are related to the multidimensional scaled association models.

Finally, to facilitate understanding and encourage future applications, readers should find it helpful that input and output files of all statistical models (written in l_{EM}, GLIM, or R) discussed in the examples can be downloaded from the Sage Publications Web site (www.sagepub.com/wongstudy). We hope that readers would find the systematic dissemination of association models and the illustrative examples useful and applicable to their own analyses involving multiway cross-classified tables.

CHAPTER 2. ASSOCIATION MODELS IN TWO-WAY TABLES

Let us begin with a two-way cross-classified table with A as the row variable and B as the column variable, with I and J categories, respectively. Note that our analysis does not require a priori distinction between dependent and independent variables, though most often empirical researchers may already have such differentiations in their analytical framework. To understand how the two (or more) variables are related to each other, whether using log-linear association or other unsaturated statistical models, it is important to recognize that they all attempt to understand the odds ratios embedded within the table under scrutiny. The major difference lies in their specific formulation of the underlying *structure* of odds ratios. By comparing the observed and expected odds ratios through their observed and expected frequencies, we can arrive at a better understanding of the relationship involved.

Odds Ratios as Fundamental Building Blocks

If we define π as the probability of success and $(1 - \pi)$ as the probability of failure, then the odds can be defined as $\Omega = \pi/(1 - \pi)$. For instance, when $\Omega = 2$, the probability of success is twice as large as the probability of failure. Conversely, when $\Omega = 0.5$, then the probability of success is only half as likely as failure. For a 2×2 table, two odds (Ω_1 and Ω_2) are available, one for each row. The ratio of the odds between Ω_1 and Ω_2 in the two rows can be defined as $\theta = \Omega_1/\Omega_2 = (\pi_{11}/\pi_{12})/(\pi_{21}/\pi_{22}) = (\pi_{11} \times \pi_{22})/(\pi_{12} \times \pi_{21})$, where π_{11}, π_{12}, π_{21}, and π_{22} represent the joint distributions of cell probabilities, with the first and second subscripts of cell probabilities representing specific row and column cells, respectively. θ is known as the odds ratio or the cross-product ratio (Yule, 1912). Note that the formulation of θ is symmetrical as its value will remain the same if the ratio of odds is derived from columns 1 and 2 instead (see Rudas, 1997, for details).

For any I by J two-way cross-classified table, there are generally $(I-1)$ $(J-1)$ unique odds ratios, although there may be different ways to calculate the complete set of odds ratios. Two strategies are common: (1) use a particular row and column (say, first row and first column or last row and last column) as reference or (2) use the adjacent rows and columns in the enumeration (Agresti, 2002, pp. 45–46). For example, if we use row i' and column j' as reference, then the logarithm of the complete set of observed odds ratios can be compactly written as follows:

$$\log\theta^*_{ij,i'j'} = \log f_{ij} + \log f_{i'j'} - \log f_{ij'} - \log f_{i'j}, \tag{2.1}$$

where f_{ij} represents the observed frequency for the (i, j) cell and so on and so forth. On the other hand, the logarithm of the observed adjacent odds ratios (or the observed local odds ratios) can be written as follows:

$$\log\theta^*_{ij} = \log f_{ij} + \log f_{i+1,j+1} - \log f_{i+1,j} - \log f_{i,j+1}, \tag{2.2}$$

where $i = 1, 2, \ldots, I - 1$ and $j = 1, 2, \ldots, J - 1$. It can be seen easily from Equations 2.1 and 2.2 that $\log\theta^*_{ij,i'j'}$ can be expressed as a function of $\log\theta^*_{ij}$, and vice versa. In other words, although the values of odds ratios under these two formulations differ, there are only $(I - 1)(J - 1)$ unique odds ratios in the two-way table. Note that the complete set of odds ratios in Equation 2.1 or 2.2 will not be affected if a particular row and/or column is multiplied by a constant. For some, this *marginal invariant* property is desirable as the odds ratios provide an understanding of the underlying association between row and column variables. For example, in the case of social mobility research, marginal discrepancies are expected because they represent the distributions of occupations or classes at different time points, and our interest focuses on relative mobility chances instead. In other social science applications, it is common to have an oversample of individuals or cases of specific attributes. The analyses of tabular data in such circumstances, however, will be unaffected by oversampling since the odds ratios remain the same in weighted or unweighted forms.

Under a particular model specification, the observed frequencies can be replaced by their corresponding expected frequencies and the complete set of the logarithm of the expected odds ratios can be represented by either log $\theta_{ij,i'j'}$ or log θ_{ij}. Note that the values of odds ratios can equal any nonnegative number, with the range between 0 and ∞. Accordingly, the range for the logarithm of the odds ratios will be between $-\infty$ and ∞. When $\theta_{ij} = 1$ or log $\theta_{ij} = 0$, it corresponds to independence between variables A and B. As noted earlier, the odds ratio can also be interpreted as the ratio between two odds. If the ratio is greater (or smaller) than 1, it means that the odds for column j versus j' for the first outcome (row i) is more likely (or less likely) than the odds for column j versus j' for the second outcome (row i'). When the ratio is 1, then the odds for both outcomes are equally likely. According to

Agresti (2002, p. 71), the standard error of $\log \hat{\theta}_{ij}$ can be calculated as follows:

$$\hat{\sigma}\left(\log \hat{\theta}_{ij}\right) = \left(\frac{1}{n_{ij}} + \frac{1}{n_{i,j+1}} + \frac{1}{n_{i+1,j}} + \frac{1}{n_{i+1,j+1}}\right)^{1/2}. \qquad (2.3)$$

Let us consider a hypothetical 2×2 table (see Table 2.1) where the row variable represents Communist party membership in China in the early 1980s (member vs. nonmember) and the column variable represents supervisory/authority position at the workplace (with subordinates vs. without subordinates). Since the odds in the first row ($40/250 = 0.16$) is less than 1, it means that even among those who were members of the Chinese Communist Party (CCP), the odds of occupying positions with authority/supervisory status as opposed to ordinary positions with no subordinates is not high. This is expected given that there should be more workers than managers/supervisors in the workplace. Similarly, one would also expect to find that the odds for nonmembers in the second row would also be less than 1 as well ($160/3000 = 0.053$), though at a much lower level than party members. Since the odds ratio or the ratio of the odds equals to 3, the number suggests that even though individual Chinese workers are more likely to be working in nonsupervisory positions, those who were members of the CCP are three times as likely as nonmembers in occupying positions with supervisory or authority status.

Independence/Null Association Model (O)

In conventional log-linear modeling of two-way tables, the baseline model that postulates independence between variables A and B is known as the independence, or null association, model (O). The model assumes nondependence or no relationship between the row (A) and the column (B) variables. Again, if we let f_{ij} and F_{ij} denote the observed and expected cell counts, respectively, of a cross-classified table for variables A and B, the model of independence or null association can be represented as the following:

$$\log F_{ij} = \lambda + \lambda_i^A + \lambda_j^B, \qquad (2.4)$$

where λ is the grand mean, λ_i^A represents the row marginal parameters, and λ_j^B represents the column marginal parameters, subject to the

Table 2.1 Hypothetical Example of a 2 × 2 Table

		Supervisory/Authority Position		
		Have Subordinates	**No Subordinates**	**Total**
A. Frequency counts				
Communist Party	Member	40	250	290
Membership	Nonmember	160	3,000	3,160
	Total	200	3,250	3,450
B. Odds				
Odds (Have	Member	40/250 = 0.160		
Subordinates vs.	Nonmember	160/3000 = 0.053		
No Subordinates)				
C. Odds ratio				
		(40)(3000)/(250)(160) = 3		

normalization that $\sum_i \lambda_i^A = \sum_j \lambda_j^B = 0$. λ_i^A and λ_j^B are marginal deviations from the grand mean, and this normalization procedure is known as effects coding. An alternative normalization is to use dummy coding where $\lambda_1^A = \lambda_1^B = 0$. Under the latter formulation, the first row and first column marginal parameters serve as the reference and values for the other λ_i^A and λ_j^B parameters represent deviations from the reference categories. The adoption of either normalization would yield identical goodness-of-fit statistics, degrees of freedom, and expected frequencies, though their individual parameter estimates differ. The model has $IJ - 1 - (I - 1) - (J - 1) = (I - 1)(J - 1)$ df. Under the model of independence, it can be shown easily that the logarithm of the (local-)odds ratios all equal to zero, that is,

$$\log\theta_{ij} = 0. \tag{2.5}$$

If the independence model is true, the log-likelihood chi-square statistic (L^2 or G^2) would distribute like a χ^2 distribution (Agresti, 2002, p. 78). One can then use the statistic to inform us whether there is significant association between variables A and B. When the independence model fails to fit the data, one can conclude that significant interaction or association between row and column variables exists, and the full interaction (FI) model would be preferred:

$$\log F_{ij} = \lambda + \lambda_i^A + \lambda_j^B + \lambda_{ij}^{AB}. \qquad (2.6)$$

Unfortunately, the FI model of Equation 2.6 uses all remaining ($I - 1$) ($J - 1$) df as interaction parameters and becomes a saturated model with 0 df. Although the interaction model might be adequate to understand the relationship between row and column variables, the number of available parameters to interpret can pose a major problem, especially when the number of row and/or column categories becomes large.

If we examine the interaction parameters carefully, we frequently observe that the number of such parameters can be significantly reduced while maintaining satisfactory fit with the data. One strategy to achieve simplification is to equate parameters with similar values and/or delete parameters that are not statistically significant. However, instead of using such an ad hoc method, it would be more desirable to develop parsimonious unsaturated models in a systematic manner that would adequately capture the interaction in the two-way cross-classified table. There is also another substantive reason why we would like to search for intermediate (unsaturated) models: Our interest often is not whether there is association between row and column variables but rather what are the underlying *patterns* or *structures* of association. Of course, we gain further confidence when the implied patterns or structures are substantively interpretable. The problem can become particularly acute in the analysis of multiple tables because while the underlying patterns or structures of association may remain the same, each table can differ from each other by their levels or scale differences.

There are many different types of models that are available to explore the patterns or structures of association; for example, topological, diagonal, and crossing models (see Goodman, 1979a, 1985; Hauser, 1978; Hout, 1983, for details). To conserve space, the current monograph will focus on only one particular family of models known as *association models* as they provide powerful and flexible ways to model underlying association pattern(s). In fact, among various statistical models that do provide satisfactory fit, association models often offer, relative to other competing ones, parsimonious and simple interpretation of the relationship involved. Furthermore, these association models can be extended easily to analyze higher-dimensional cross-classification tables (see later chapters for details) that are particularly useful in social science applications.

One-Dimensional Association Models

The family of association models can be classified along two specifications: functional forms and dimensionality or degree of complexity. The former can take log-linear, log-multiplicative, or hybrid formulations, whereas the latter can be one-dimensional, two-dimensional, or multidimensional. Of course, hybrid formulations that combine both log-linear and log-multiplicative specifications are by definition at least two-dimensional. Although association models are often termed as ordinal models, they can be applied to cross-classified tables that are either nominal or ordinal. In other words, the family of association models discussed in the entire monograph can be applied more generally to nominal-nominal, nominal-ordinal, and ordinal-ordinal variables. On the other hand, association models can become truly ordinal models if one also assumes a monotone relationship; that is, the (observed or estimated) row and/or column scores are monotonically increasing or decreasing. This can be achieved by imposing order or inequality restrictions so that all observed or estimated row and/or column scores are at the proper order (Agresti & Chuang, 1986; Agresti, Chuang, & Kezouh, 1987; Bartolucci & Forcina, 2002; Galindo-Garre & Vermunt, 2004; Ritov & Gilula, 1991). The imposition of such restrictions will be illustrated in some of the examples later.

To understand major differences between various formulations of association models, it is important to recognize that they differ from each other in terms of whether there are specified/unspecified ordering of categories and specified/unspecified spacing between categories (Goodman, 1985; see Table 2.2 for details). For instance, if we have both specified ordering and specified spacing of categories for both row and column variables, the most parsimonious model to describe the underlying association would be the uniform association model (U), although the log-linear row and column effects model ($R+C$), the row effects model (R), and the column effects model (C) would be appropriate too.

On the other hand, if we have both unspecified order and unspecified spacing of categories, then the log-multiplicative row and column effects (RC) model is the only appropriate model. As indicated in Table 2.2, different combinations of these two conditions will lead to different model specifications. If several models provide satisfactory results, we have to weigh between the relative importance of model accuracy and scientific parsimony in choosing the final model to understand the association involved. We will discuss some common strategies in the model selection section later. Furthermore, because of the relationship displayed in the table here, one can understand why these models are members of a family of association models and their systematic relationships among each other.

Table 2.2 Assumptions on Ordering and Spacing Between Categories in
Association Models

Models	Unspecified Order	Specified Order	
		Unspecified Spacing	Specified Spacing
0	Rows and columns	—	—
U	—	—	Rows and columns
R	Rows	Rows	Columns
C	Columns	Columns	Rows
$R+C$	—	—	Rows and columns
RC	Rows and/or columns	Rows and/or columns	—

SOURCE: Adapted from Goodman (1985, Table 4A).

Uniform Association (U) Model

When the ordering and spacing of categories for both row and column
variables are known or specified, it is possible to postulate an association
model that uses only 1 *df* to describe the association involved. This is
known as the uniform association model (U) (Duncan, 1979; Goodman,
1979b). Suppose U_i and V_j are fixed integer scores for the row (A) and column (B) variables, say $U_i = 1, \ldots, I$ and $V_j = 1, \ldots, J$. The uniform association model (U) can be written formally as follows:

$$\log F_{ij} = \lambda + \lambda_i^A + \lambda_j^B + \beta U_i V_j, \qquad (2.7)$$

where β is the uniform association parameter. Note that the model is delib-
erately written in this form to demonstrate its relationship with other association models discussed later. When a priori scores are used, Equation 2.7
is instead known as the generalized uniform association (U^o) model (see
Goodman, 1986, 1991; Hout, 1983, for details), which is equivalent to
Haberman's (1978) linear-by-linear association. The model has $(I - 1)$
$(J - 1) - 1 = IJ - I - J$ *df*. For the sake of simplicity, let us assume here that
fixed integer scores are used instead, then the adjacent odds ratios can be
simplified as the following:

$$\log \theta_{ij} = \beta(U_{i+1} - U_i)(V_{j+1} - V_j) = \beta, \qquad (2.8)$$

because $U_{i+1} - U_i = V_{j+1} - V_j = 1$. Under the U model, the consecutive rows are equidistant, and the consecutive columns are also equidistant. Note that if U_i and V_j are not fixed integer scores, the above relationship in Equation 2.8 cannot be simplified. In either case, if either the U or U^ρ model fits the data well, the complete set of adjacent odds ratios can be captured by just one single parameter, β. Of all association models discussed in this monograph, the U model is therefore the most parsimonious as well as the most restrictive one.

Row Effects (R) Model

When only the column variable has known ordering and spacing that can be represented by fixed integer scores, Equation 2.7 becomes the row effects model (R). Algebraically, the row effects model can be written as follows:

$$\log F_{ij} = \lambda + \lambda_i^A + \lambda_j^B + \tau_i^A V_j. \tag{2.9}$$

Note that only $(I - 1)$ τ_i^A parameters can be identified. To uniquely identify all τ_i^A parameters, one can impose the restriction that either $\tau_1^A = 0$ or $\sum \tau_i^A = 0$. Therefore, the R model has $(I-1)(J-1) - (I-1) = (I-1)(J-2)$ df. The reason why the model is termed as the row effects model is because all adjacent odds ratios involving row i as opposed to row i' will be identical, and one can rank order the row effects parameters to indicate differences in strength of association. The latter interpretation can be better understood as the adjacent odds ratios can be represented as follows:

$$\log \theta_{ij} = \left(\tau_{i+1}^A - \tau_i^A\right)\left(V_{j+1} - V_j\right) = \tau_{i+1}^A - \tau_i^A. \tag{2.10}$$

Column Effects (C) Model

When only the row variable has known ordering and spacing that can be represented by fixed integer scores (U_j), then the column effects (C) model can be written as the following:

$$\log F_{ij} = \lambda + \lambda_i^A + \lambda_j^B + \tau_j^B U_i. \tag{2.11}$$

Again, only $(J-1)$ τ_j^B parameters can be uniquely identified. To identify all τ_j^B parameters, one can impose the restriction that either $\tau_1^B = 0$ or $\sum \tau_j^B = 0$. The column effects (C) model has $(I-1)(J-1) - (J-1) = (I-2)(J-1)$ df. Under the C model, all adjacent odds ratios involving column j as opposed to column j' will be identical because the adjacent odds ratios can be written as follows:

$$\log\theta_{ij} = \left(\tau_{j+1}^{B} - \tau_{j}^{B}\right)(U_{i+1} - U_i) = \tau_{j+1}^{B} - \tau_{j}^{B}. \qquad (2.12)$$

Log-Linear Row and Column Effects (R+C) Model

To take full advantage of the specified spacing and ordering of row and column variables simultaneously, one can specify a model that calculates both row effects and column effects by using the specified spacing of column and row variables. The statistical model that uses such specification is known as the log-linear row and column effects model (*R+C*) because the row and column effects are log additive. When this model was initially formulated by Goodman (1979b), it was known as the *RC* association *I* model, in contradistinction from another *RC* association *II* model that is log multiplicative in nature. Such semantic distinction was dropped from his later works (Goodman, 1985, 1986, 1991), and they are now commonly distinguished as *R+C* and *RC* models instead. The latter distinction has remained popular among empirical practitioners since then. Algebraically, the log-linear row and column effects model (*R+C*) can be written as follows:

$$\log F_{ij} = \lambda + \lambda_i^A + \lambda_j^B + \tau_i^A V_j + \tau_j^B U_i. \qquad (2.13)$$

Note that unlike the row effects model (*R*) and the column effects model (*C*) discussed earlier, there are only (a) $(I-1)\tau_i^A$ and $(J-2)\tau_j^B$ or (b) $(I-2)\tau_i^A$ and $(J-1)\tau_j^B$ parameters under Equation 2.13 that can be simultaneously and uniquely identified. In the former case, one can impose the restriction that $\tau_1^A = \tau_1^B = \tau_j^B = 0$. Instead of $\tau_1^A = 0$, the normalization that $\sum \tau_i^A = 0$ is also possible. In the latter case, the normalization that $\tau_1^A = \tau_I^A = \tau_1^B = 0$ can also uniquely identify all parameters. Similarly, instead of $\tau_1^B = 0$, it is possible to adopt the normalization that $\sum \tau_j^B = 0$ as well. In sum, the *R+C* model uses an additional $I + J - 3$ parameter than the independence model and has $(I-2)(J-2)$ *df*.

Finally, it is also possible to rewrite the *R+C* model as follows:

$$\log F_{ij} = \lambda + \lambda_i^A + \lambda_j^B + \beta U_i V_j + \tau_i^A V_j + \tau_j^B U_i, \qquad (2.14)$$

To uniquely identify all row and column effects parameters under Equation 2.14, the normalization that $\tau_1^A = \tau_I^A = \tau_1^B = \tau_J^B = 0$ is necessary. Written in this form, it is evident that *U*, *R*, *C*, and *R+C* models are all related to each other.

Based on Equation 2.14, the logarithm of the adjacent odds ratios under the *R*+*C* model can be written as follows:

$$\log\theta_{ij} = \beta(U_{i+1} - U_i)(V_{j+1} - V_j) + (\tau_{i+1}^A - \tau_i^A)(V_{j+1} - V_j)$$
$$+ (\tau_{j+1}^B - \tau_j^B)(U_{i+1} - U_i) = \beta + (\tau_{i+1}^A - \tau_i^A) + (\tau_{j+1}^B - \tau_j^B). \quad (2.15)$$

When there is a one-to-one correspondence between row and column variables, that is, a squared table with equal number of row and column categories as in the case of the study of marriage homogamy and intergenerational mobility, one can impose an even more parsimonious model that equates the row effects parameters (τ_i^A) with the column effects parameters (τ_j^B) for the same categories. The latter is known as the equal log-linear row and column effects (equal *R*+*C*) model and has $(I-1)(I-2)$ *df* since $I=J$. The equal *R*+*C* model is a special case of the model of quasi-symmetry (*QS*) because the underlying association will be symmetrical in nature. By the same token, the uniform association (*U*) model for a squared table with one-to-one correspondence between row and column categories would imply *QS* as well.

Log-Multiplicative Row and Column Effects (RC) Model

Unlike previous association models that make explicit assumptions about specified ordering and spacing of row and/or column categories, the last one-dimensional association model in this section relaxes such assumption. Instead, the model attempts to derive both row and column scores empirically from the association pattern found in the cross-classified table. It assumes that the row and column score parameters are related to each other in log-bilinear or log-multiplicative form (Andersen, 1980, 1991; Goodman, 1979b, 1981b, 1985; Haberman, 1981). Algebraically, the *RC* model can be represented as follows:

$$\log F_{ij} = \lambda + \lambda_i^A + \lambda_j^B + \phi\mu_i\nu_j, \quad (2.16)$$

subject to the normalization constraints that $\sum_i \mu_i = \sum_j \nu_j = 0$ and $\sum_i \mu_i^2 = \sum_j \nu_j^2 = 1$ (Goodman, 1979b). These constraints are otherwise known as the centering and scaling restrictions, respectively. Similar to the *R*+*C* model, the *RC* model has $(I-2)(J-2)$ *df*.

The above constraints produce *unweighted* or *unit standardized solutions* for the row score parameters (μ_i) and column score parameters (ν_j). Other normalized constraints can also be used to identify the model. For example,

Goodman (1981b) has proposed to weigh the row and column scores with respect to their marginal weights to relate the *RC* model to the canonical correlation approach. That is, $\sum_i \mu_i P_{i\cdot} = \sum_j \nu_j P_{\cdot j} = 0$ and $\sum_i \mu_i^2 P_{i\cdot} = \sum_j \nu_j^2 P_{\cdot j} = 1$ where $P_{i\cdot}$ and $P_{\cdot j}$ are the row and column marginal probabilities, respectively, and the resultant row and column score parameters represent *marginal-weighted solutions*. It is also possible to impose another set of constraints: $\mu_1 = 1$, $\mu_I = I$, $\nu_1 = 1$, and $\nu_J = J$ to identify the *RC* model. Note that for single-table analysis, the adoption of different constraints will not affect the interpretation of results in any significant way. However, in analyzing multiway cross-classification tables, especially those involving grouping variable(s), the choice of different weights may lead to radically different results. Following the recommendations by Clogg and his associates (Becker & Clogg, 1989; Clogg & Rao, 1991; Clogg & Shihadeh, 1994), it is more preferable to adopt the unit standardized weights to facilitate comparison.

The row and column (μ_i and ν_j) scores in the *RC* model can be viewed as "bivariate normal scores" and/or in terms of maximizing the intrinsic association between row and column variables. Because of the latter interpretation, ϕ is known as the *intrinsic* association parameter and denotes the strength of association when the row and column score parameters are both one unit in length. Unlike other statistical measures such as the Pearson correlation (r), the value of ϕ is always greater than 0 and unbounded (i.e., $0 < \phi < \infty$).[1] Perhaps, the most important property of the *RC* model is that any interchange(s) of rows and/or columns will not affect the values of estimated score parameters. This desirable property means that researchers can obtain rankings and distances of categories a posteriori when they are unsure about the exact rankings and distances between categories. Similarly, when there is a one-to-one correspondence between row and column variables, one can also estimate the equal log-multiplicative row and column effects model by imposing that $\mu_i = \nu_j$ for all $i = j$. Similar to the equal *R+C* model, the equal *RC* model has $(I - 1)(I - 2)$ *df* and is a special case of the *QS* model.

It should be noted that the intrinsic ordering of the categories of each variable is determined by an ordering in the pattern of the *joint* distribution of the variables of interests. The exact ordering depends on the introduction of relevant or appropriate variable(s). Accordingly, it may be more appropriate to denote the "intrinsic" ordering as "extrinsic" or "contingent" ordering instead (Goodman, 1987, p. 530), as the choice of different variables as the column variable, for example, may yield different orderings of the row variable, and vice versa.

It is also possible to rewrite Equation 2.16 as the following:

$$\log F_{ij} = \lambda + \lambda_i^A + \lambda_j^B + \mu_i^* \nu_j^*, \tag{2.17}$$

where $\mu_i^* = \phi^\gamma \mu_i$ and $\nu_j^* = \phi^\delta \nu_j$ for any γ and δ as long as their sum equals to 1, that is,

$$\gamma + \delta = 1. \tag{2.18}$$

Note that only one scaling constraint for either μ_i^* or ν_j^* is needed under the reformulation. When γ and δ are both equal to 0.5, μ_i^* and ν_j^* are known as the adjusted row and column scores, respectively, and they are routinely reported in CDAS and l_{EM}. The respecification provides useful ways to present RC-type models in graphical displays (see later examples for details).

The structure of the expected adjacent log-odds ratios under the RC model is as follows:

$$\log \theta_{ij} = \phi(\mu_{i+1} - \mu_i)(\nu_{j+1} - \nu_j). \tag{2.19}$$

Comparing with previous association models, the implied structure of expected odds ratios is more complex. Because μ_i and ν_j are both parameters, the product term in Equation 2.19 cannot be further simplified. The major distinction between $[U_i, V_j]$ in the R+C model and $[\mu_i, \nu_j]$ in the RC model is that the former uses fixed assigned (integer) scores, whereas the latter adopts score parameters estimated from empirical data.

Because the RC model is log-multiplicative in nature, the cyclic application of Newton's unidimensional algorithm (Becker, 1990; Clogg, 1982a; Goodman, 1979b) has been proposed to estimate all association parameters (ϕ, μ_i, and ν_j). With each iterative cycle, the procedure estimates one set of parameters (say, μ_i) while treating the others (say, ϕ and ν_j) as fixed scores. The procedure will yield maximum likelihood estimates when the difference in estimates (likelihood ratio test [LRT] statistics or parameter estimates) from the current and previous cycles is smaller than a prespecified, small convergence criterion. With the recent introduction of the *gnm* package in R to estimate generalized nonlinear models (Turner & Firth, 2007a, 2007b), it is now possible to estimate both association parameters and their asymptotic standard errors simultaneously via the modified or stabilized Newton-Raphson algorithm (see also Aït-Sidi-Allal, Baccini, & Mondot, 2004; Gilula & Haberman, 1986; Haberman, 1979, 1995). Unless otherwise stated, the standard errors of all association parameters reported in this monograph are obtained from R directly. Their parameter estimates are initially estimated from another package, l_{EM}, and have been verified by R. If the standard errors cannot be obtained directly because of the current limitation of the *gnm* module, then the bootstrap standard errors are reported instead.

From the above discussion, one can see that there are systematic relationships between O, U, R, C, $R+C$, and RC models (and other more complicated association models in the next section). According to Goodman (1981a, 1981b, 1985, 1991), the RC association model can be treated as an approximation to the discretized bivariate normal distribution, whereas the U association model can be used when the discretized bivariate normal has equal-length row intervals and equal-length column intervals, except for the first- and last-row intervals and the first- and last-column intervals. This interpretation is particularly true when marginal weights are used instead of unweighted association model.

Assuming that there is no need to entertain two- or higher-dimensional association models, the general modeling strategy is to compare the goodness-of-fit statistics of the above models relative to their degrees of freedom. According to the procedure of the LRT, the contrast between the goodness-of-fit statistics of two (nested) models will be asymptotically χ^2 distributed with degrees of freedom equal to the difference between the two models' degrees of freedom. Also, through the partitioning of the chi-square statistics, empirical researchers can create an analysis of association (ANOAS) table to understand the relative contribution of each component (uniform association, row effects, column effects, and row and column effects). Details about this partitioning strategy can be seen in the illustrative examples throughout this monograph. Interested readers may want to consult the works of Goodman (1979b, 1981a) and Clogg and Shihadeh (1994) for similar partitioning strategies.

Two-Dimensional Association Models

If none of the above one-dimensional association models fits the data well, we can increase model complexity by incorporating additional interaction terms. Two options are generally available within the association model framework. Since the association structure can be captured by either log-linear or log-multiplicative components, one option is to combine both components together, resulting in four specific two-dimensional association models ($U+RC$, $R+RC$, $C+RC$, and $R+C+RC$). These are *hybrid* models, though they differ from those introduced by Wong (1990, 1992); the latter incorporates both vertical and nonvertical effects in examining the association between row and column variables. Another option simply adds another dimension to the *RC* model, resulting in the *RC*(2) model. It should be stressed that the choice between them is empirical as well as theoretical and depends on their relative goodness of fit, ease of understanding, and substantive interpretation.

U+RC Model

The first two-dimensional association model combines two components: uniform association (*U*) and log-multiplicative row and column effects (*RC*) into the same model. The resulting *U+RC* model can be written as follows:

$$\log F_{ij} = \lambda + \lambda_i^A + \lambda_j^B + \phi_1 U_i V_j + \phi_2 \mu_i \nu_j. \tag{2.20}$$

Note that the uniform association parameter is now written as ϕ_1 instead of β_1 and the intrinsic association in the *RC* component as ϕ_2 to highlight that this is a two-dimensional association model. As usual, both centering and scaling constraints are needed to uniquely identify μ_i and ν_j. The model has $IJ - 2I - 2J + 3$ *df*, and the logarithm of the adjacent odds ratios can be shown to consist of two components:

$$\log \theta_{ij} = \phi_1 + \phi_2 (\mu_{i+1} - \mu_i)(\nu_{j+1} - \nu_j). \tag{2.21}$$

One can compare the goodness-of-fit statistic from *U+RC* model with those of *U* and/or *RC* models to evaluate whether the additional dimension of association is indeed warranted. The comparison will also allow us to scrutinize the relative contributions of each component in Equation 2.20 and assess whether the principal dimension of association is log-linear or log-multiplicative in nature.

R+RC Model

Instead of using a single uniform association parameter, one can substitute it with the row effects parameters (*R*), with fixed integer column scores (V_j). This would result in a slightly more complex hybrid model, and the *R+RC* model can be written as follows:

$$\log F_{ij} = \lambda + \lambda_i^A + \lambda_j^B + \tau_i^A V_j + \phi_2 \mu_i \nu_j, \tag{2.22}$$

where τ_i^A represents the row effect parameters. Unlike its one-dimensional counterpart, we need two constraints to identify all τ_i^A parameters (e.g., $\tau_1^A = \tau_I^A = 0$). On the other hand, the same centering and scaling constraints are required to identify both μ_i and v_j. Therefore, the model has $(I-2)(J-2) - (I-2) = (I-2)(J-3)$ *df*. By comparing the goodness-of-fit statistic of the present model with its simpler one-dimensional counterparts, one can assess whether the more complex formulation is indeed consistent with the data. When both components contribute significantly to our understanding, the complex model should be preferred over the simpler but erroneous formulation. Finally, the logarithms of the adjacent odds ratios can be found to consist of two (log-linear and log-multiplicative) components:

$$\log\theta_{ij} = \left(\tau_{i+1}^A - \tau_i^A\right) + \phi_2(\mu_{i+1} - \mu_i)(v_{j+1} - v_j). \tag{2.23}$$

C+RC Model

Instead of assuming the existence of row effects in Equation 2.22, one can substitute them with column effects (τ_j^B) and fixed integer row scores (U_i). The resulting model becomes the *C+RC* model, that is,

$$\log F_{ij} = \lambda + \lambda_i^A + \lambda_j^B + \tau_j^B U_i + \phi_2\mu_i v_j. \tag{2.24}$$

Again, two constraints are needed to identify all τ_j^B parameters (say, $\tau_1^B = \tau_J^B = 0$), and the model has $(I-2)(J-2) - (J-2) = (I-3)(J-2)$ *df*. Similarly, the logarithms of the adjacent odds ratios consist of both log-linear and log-multiplicative components, that is,

$$\log\theta_{ij} = \left(\tau_{j+1}^B - \tau_j^B\right) + \phi_2(\mu_{i+1} - \mu_i)(v_{j+1} - v_j). \tag{2.25}$$

The relative contribution of each component can be compared with its lower-order counterparts by using the ANOAS partitioning strategy.

R+C+RC Model

Another hybrid formulation incorporates both log-linear row and column effects (*R+C*) and log-multiplicative row and column effects (*RC*) together. The resultant *R+C+RC* model can be written as follows:

$$\log F_{ij} = \lambda + \lambda_i^A + \lambda_j^B + \phi_1 U_i V_j + \tau_i^A V_j + \tau_j^B U_i + \phi_2\mu_i v_j. \tag{2.26}$$

To uniquely identify the log-linear row and column effects parameters, one additional constraint on one of the τ_i^A or τ_j^B parameters is needed; for example, $\tau_1^A = \tau_2^A = \tau_I^A = \tau_1^B = \tau_J^B = 0$ or $\tau_1^A = \tau_I^A = \tau_1^B = \tau_2^B = \tau_J^B = 0$. Therefore, the model has $(I-1)(J-1) - (I+J-5) - (I-2) - (J-2) - 1 =$

$(I-3)(J-3)$ *df.* The logarithms of the adjacent odds ratios have the following structure:

$$\log\theta_{ij} = \phi_1 + \left(\tau^A_{i+1} - \tau^A_i\right) + \left(\tau^B_{j+1} - \tau^B_j\right)$$
$$+ \phi_2(\mu_{i+1} - \mu_i)(v_{j+1} - v_j). \tag{2.27}$$

RC(2) Model

The last two-dimensional association model drops the log-linear term and replaces it with another log-multiplicative component. It is denoted as the *RC*(2) model, where the value in the bracket represents the number of dimensions of log-multiplicative components. It can be represented as follows:

$$\log F_{ij} = \lambda + \lambda^A_i + \lambda^B_j + \phi_1\mu_{i1}v_{j1} + \phi_2\mu_{i2}v_{j2}, \tag{2.28}$$

and the logarithms of adjacent odds ratios have the following structure:

$$\log\theta_{ij} = \phi_1\left(\mu_{i+1,1} - \mu_{i1}\right)\left(v_{j+1,1} - v_{j1}\right)$$
$$+ \phi_2\left(\mu_{i+1,2} - \mu_{i2}\right)\left(v_{j+1,2} - v_{j2}\right). \tag{2.29}$$

Similar to the *R+C+RC* model, it also has $(I-3)(J-3)$ *df.* In addition to the centering and scaling constraints for both row and column score parameters (μ_{i1}, μ_{i2}, v_{j1}, and v_{j2}), that is, they all have a sum of zero and a sum of squares of 1, we need additional cross-dimensional constraints to uniquely identify all parameters. They are as follows:

$$\sum\nolimits_{i=1}^{I} \mu_{i1}\mu_{i2} = \sum\nolimits_{j=1}^{J} v_{j1}v_{j2} = 0 \tag{2.30}$$

In other words, the vectors μ_{i1} and μ_{i2} form an orthonormal basis and so do v_{j1} and v_{j2}. If there is a one-to-one correspondence between row and column categories, it may be interesting to impose equality constraints on some if not all of the row and column scores in both dimensions, for example, $\mu_{i1} = v_{j1}$ for all $i = j$ in the first dimension only. Of course, when equal row and column scores are imposed on both dimensions, the underlying association pattern will be symmetrical and the model of *QS* is implied.

If two sets of a priori fixed row and column scores $[U_{i1}, V_{j1}]$ and $[U_{i2}, V_{j2}]$ are used in place of $[\mu_{i1}, v_{j1}]$ and $[\mu_{i2}, v_{j2}]$, the model can be denoted as $U^o_1 + U^o_2$. This bears a strong resemblance with the SAT (status, autonomy, and training) model developed by Hout (1984), except that his formulation is even more complex. The latter is actually three- and higher-dimensional and includes additional effects for diagonal cells. Based on the exposition presented so far, it is obvious that $U^o_1 + U^o_2$ is indeed a special case of the *RC*(2) model or the *RC*(*M*) model that will be developed shortly in the next section. Because the $RC(2)$ model requires cross-dimensional constraints, some readers may prefer to use hybrid models instead because the latter can be readily estimated from standard statistical packages when only a handful of statistical packages have built-in options for cross-dimensional constraints. Nonetheless, the $RC(2)$ model can still be attractive when we do not want to impose any a priori ordering of categories but are interested to obtain score estimates a posteriori. Furthermore, as it will be illustrated in Chapter 4 for multiway cross-classification tables, it is possible that, under special circumstances, there is no need to impose cross-dimensional constraints for some $RC(2)$ or related models with complex specifications.

Multidimensional $RC(M)$ Association Model

Although it is possible to formulate hybrid models in three- or higher-dimensional association models, it is perhaps more insightful to generalize the $RC(2)$ association model to the Mth dimension instead (Clogg & Shihadeh, 1994; Goodman, 1985).[2] Generally speaking, the multidimensional $RC(M)$ association model can be denoted as follows:

$$\log F_{ij} = \lambda + \lambda_i^A + \lambda_j^B + \sum_{m=1}^{M} \phi_m \mu_{im} \nu_{jm}, \tag{2.31}$$

where $0 \le M \le \min(I-1, J-1)$. For any $M^* < M$, the model becomes unsaturated and has $(I - M - 1)(J - M - 1)$ df. When $M^* = \min(I, J) - 1$, it becomes the saturated or the FI model. When $M^* = 0$, it is equivalent to the independence or null association model (O). To identify all row and column scores, parameters, centering, scaling, and cross-dimensional constraints are needed. For instance, the following cross-dimensional constraints are needed:

$$\sum_{i=1}^{I} \mu_{im} \mu_{im'} = \sum_{j=1}^{J} \nu_{jm} \nu_{jm'} = 0, \tag{2.32}$$

where $m \ne m'$. Alternatively, the scaling and cross-dimensional constraints can be written more compactly as follows:

$$\sum_{i=1}^{I} \mu_{im} \mu_{im'} = \sum_{j=1}^{J} \nu_{jm} \nu_{jm'} = \delta_{mm'}, \tag{2.33}$$

where $\delta_{mm'}$ is the Kronecker δ (Becker, 1990; Becker & Clogg, 1989; Goodman, 1985). The complete set of adjacent log-odds ratios under this specification has the following structure:

$$\log \theta_{ij} = \sum_{m=1}^{M} \phi_m (\mu_{i+1,m} - \mu_{im})(\nu_{j+1,m} - \nu_{jm}). \tag{2.34}$$

Without loss of generality, we can rearrange the intrinsic association parameters such that $\phi_1 \ge \phi_2 \ge \ldots \ge \phi_M \ge 0$.

Alternatively, one can rewrite Equation 2.31 as the following:

$$\log F_{ij} = \lambda + \lambda_i^A + \lambda_j^B + \sum_{m=1}^{M} \mu_{im}^* v_{jm}^*, \qquad (2.35)$$

where $\mu_{im}^* = \phi_m^\gamma \mu_{im}$, $v_{jm}^* = \phi_m^\delta v_{jm}$, and $\gamma + \delta = 1$. There are at least three possible normalizations that one can adopt to identify μ_{im}^* and v_{jm}^* in Equation 2.35. They are as follows:

(a) Row principal normalization

$$\mu_{im}^* = \phi_m \mu_{im} \text{ and } v_{jm}^* = v_{jm}. \qquad (2.36)$$

(b) Column principal normalization

$$\mu_{im}^* = \mu_{im} \text{ and } v_{jm}^* = \phi_m v_{jm}. \qquad (2.37)$$

(c) Symmetrical normalization

$$\mu_{im}^* = \sqrt{\phi_m} \mu_{im} \text{ and } v_{jm}^* = \sqrt{\phi_m} v_{jm}. \qquad (2.38)$$

For most practical purposes, the last option is usually preferred because it does not make any preference to either rows or columns in normalization (Clogg & Shihadeh, 1994). Furthermore, the estimated row and column scores from symmetrical normalization will be helpful in graphical displays that would assist our understanding on how estimated row and/or column scores from different dimensions are related to each other.[3]

Relationship Among Various Association Models

Based on the above exposition, one may already recognize that the U, R, C, $R+C$, RC, and their higher-order counterparts have systematic relationships with each other. For example, one can rewrite the row effects parameters (τ_i^A) under the row effects (R) model in Equation 2.9 as the following:

$$\tau_i^A = \zeta^* \zeta_{i.}^*, \qquad (2.39)$$

by factoring out an overall effect, with the restrictions that

$$\sum_{i=1}^{I} \zeta_{i.}^* = 0 \text{ and } \sum_{i=1}^{I} \zeta_{i.}^{*2} = k, \qquad (2.40)$$

where k is any arbitrary constant, say 1 or I. Under this respecification, there are still $(I-1)$ nonredundant parameters: $(I-2)$ for ζ_i^* and 1 for ζ^*. The uniform association (U) model therefore can be seen as a special case of R when $\zeta_i^* = 1$ for all i. Similarly, the column effects (C) model can also be reexpressed to include two components: ζ^* and ζ_j^*, and the uniform association model is also a special case of the C model as well. On the other hand, the R and C models do not have any systematic relationship with each other. Furthermore, by comparing Equations 2.9, 2.11, 2.13, and 2.14, one can also understand why the R and C models can be considered as special cases of the $R+C$ model.

What about the relationship between R and RC models? Suppose we let $v_j = V_j$, that is, with fixed integer scores, then the RC model can be written as follows:

$$\log F_{ij} = \lambda + \lambda_i^A + \lambda_j^B + \phi\mu_i V_j = \lambda + \lambda_i^A + \lambda_j^B + V_i^* \tau_i^{A*}, \quad (2.41)$$

where $V_i^* = V_j$ and $\tau_i^{A*} = \phi\mu_i$. Under this expression, the R model becomes a special case of the RC model when consecutive columns are equidistant. By the same token, the C model is a special case of the RC model when the consecutive rows are equidistant. Finally, the U model is a special case of the RC model when the consecutive rows and columns are both equidistant. On the other hand, there is no direct relationship between the $R+C$ model and the RC model.

Systematic relationships between the family of association models can be extended to higher dimensions as well. For example, if the row and column scores in the first dimension of the $RC(M)$ model can be expressed as fixed integer scores, that is, $\mu_{i1} = U_{i1}$ and $v_{j1} = V_{j1}$, then the model is equivalent to the $U+RC(m^*)$ model, where $m^* = 1, \ldots, M-1$. By the same token, one can reexpress the $RC(M)$ model as the $R+RC(m^*)$ model when only $v_{j1} = V_{j1}$, as the $C+RC(m^*)$ model when only $\mu_{i1} = U_{i1}$, and as the $R+C+RC(m^*)$ when $\mu_{i1} = U_{i1}$ and $v_{j1} = V_{j1}$ but their effects are in additive forms instead. Finally, if M set of a priori scores is used for both row and column categories, the $RC(M)$ model can be denoted as $U^p(m)$, where $m = 1, 2, \ldots, M$ or $U^p(1)+\cdots+U^p(M)$ (Hout, 1984, 1988). For illustrative purposes, Figure 2.1 presents a graphical relationship between various one-dimensional and two-dimensional association models (see also Goodman, 1985).

Model Estimation, Degrees of Freedom, and Model Selection

Several statistical packages are available to estimate association models. They include CDAS 3.50 (Eliason, 1990), 1_{EM} 1.0 (Vermunt, 1997), GLIM 4.09

24

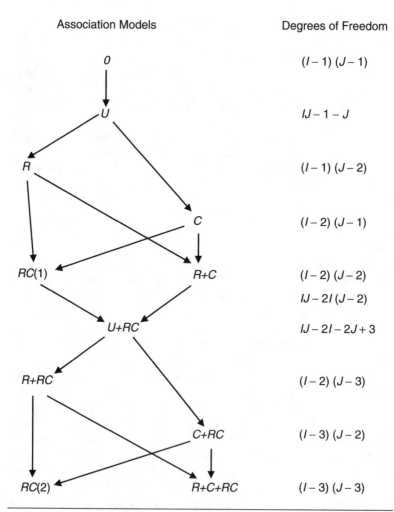

Association Models

Degrees of Freedom

0          $(I-1)(J-1)$

U $IJ-1-J$

R $(I-1)(J-2)$

C $(I-2)(J-1)$

RC(1)   R+C $(I-2)(J-2)$
                             $IJ-2I(J-2)$

U+RC $IJ-2I-2J+3$

R+RC $(I-2)(J-3)$

C+RC $(I-3)(J-2)$

RC(2)   R+C+RC $(I-3)(J-3)$

Figure 2.1   Graphical Relationship Between Some Association Models for the $I \times J$ Table

SOURCE: Adapted from Goodman (1985, Figure 1).

NOTE: The calculation of degrees of freedom may not be applicable when $I = 2$ and/or $J = 2$.

(Francis, Green, & Payne, 1993), and R, particularly the *gnm* package (Firth & Menezes, 2004; Ihaka & Gentleman, 1996; Turner & Firth, 2007a, 2007b). In particular, CDAS, 1_{EM}, and R are freewares that can be downloaded and installed in most personal computers.[4] Some of them can also be estimated via user-written standalone programs or specialized modules in SAS or STATA as well. All models presented in the illustrative examples here are estimated from either l_{EM}, GLIM, or R, and their input and output files can be downloaded from the Sage Publications Web site (www.sagepub.com/wongstudy).

Generally speaking, l_{EM} is the preferred statistical package because of its speed and ease of use. However, since the current version of l_{EM} does not impose cross-dimensional constraints for multidimensional $RC(M)$ association models, the latter models have to be estimated via a user-written iterative macro program from GLIM that imposes the Gram-Schmidt orthonormalization restrictions (Wong, 2001). It is also possible to estimate the RC model from STATA by the RC2 and MCLEST ado programs written by John Hendrickx and $RC(M)$ models from the standalone DASSOC program written by Haberman (1995).[5] Unfortunately, these user-written programs have one major limitation: They are restricted to two-way tables and cannot be extended to multiway tables. Fortunately, most of the estimation issues can now be resolved because of the recent addition of the generalized nonlinear models (*gnm*) module by Turner and Firth (2007a, 2007b) in R.[6] It is likely that the latter will become the default statistical package in the future because of its flexibility to impose various types of restrictions and obtain asymptotic standard errors at the same time.

One major problem in estimating RC-type association models is that the likelihood function may occasionally have multiple local maxima. This is likely to occur for models with poor fit and complex restrictions, particularly those involving multiple tables. In these circumstances, the postulated models may not converge properly. Two general strategies can be adopted to resolve this particular problem. The first strategy involves the use of multiple random start values to ensure that the parameter estimates all converge to the same values (down to the fourth digits, for instance). The second one sets the default convergence criterion (e.g., 0.000001 in l_{EM}) to an even smaller constant, say 0.0000000001 instead. Another problem is that the algorithm may stop after a few iterations because of a decrease of likelihood. This usually occurs because of poor starting values or because the model is far too complex to be estimated properly. While the former can sometimes be resolved by using random starting values, it is perhaps best to resolve the latter problem through model simplifications instead.

Generally speaking, the calculation of the degrees of freedom of a model depends on the total number of cells minus the number of uniquely identified parameters of a particular model. However, the accounting of the latter can become problematic for multidimensional association models because of the existence or the lack of cross-dimensional constraints. As we shall see in the next chapter, problems may arise for certain $RC(M)$ models where no or only some but not all possible cross-dimensional constraints are needed for a number of more restricted models. We will provide a more thorough discussion of those situations in Chapters 3 and 4. In any event, if we know the number of cross-dimensional constraints imposed, the degrees of freedom can be calculated as follows:

df = Number of cells − (Number of parameters − Number of restrictions)
   = Number of cells − Number of parameters + Number of restrictions.

It is important to stress that knowledge about the calculation of degrees of freedom of a model is critical because the reported figures from available statistical packages such as CDAS and l_{EM} for some models are incorrect. On the other hand, R can provide the correct degrees of freedom in most cases under appropriate conditions.

Through the estimation of a series of association models and/or other competing models, it is necessary to develop a systematic strategy in model selection that would best describe the underlying association patterns. In theory, two factors should be considered: model accuracy and scientific parsimony. Everything else being equal, the Occam's razor principle should be adopted (Kotz & Johnson, 1985, pp. 578–579). Suppose E represents the evidence and $p(H|E)$ represents the probability of a specified hypothesis H given the evidence of E. The principle states that if

$$p(H_1|E) = p(H_2|E) = \ldots = p(H_k|E), \qquad (2.42)$$

for hypotheses $H_1, \ldots, H_k$, then the simplest one among $H_1, \ldots, H_k$ is to be preferred. In practice, however, model accuracy and scientific parsimony have often been construed as trade-offs, especially when the sample size becomes large. For instance, it is not uncommon for researchers to offer statements such as "With small sample sizes we can often find models to give a satisfactory fit; with large samples, no model fits" (Goodman, 1991, p. 1085; see also Berkson, 1938; Diaconis & Efron, 1985) or "Our results [with large samples] are purely negative; no matter what model we try, we are sure to find significant deviations which force us to reject [the model]" (Goodman, 1991, p. 1085; see also Fisher, 1925; Martin-Löf, 1974).

There are many different strategies that empirical researchers can adopt to deal with the problem of relatively poor fitting models with large samples. Among them, the Bayesian information criterion (BIC) statistic perhaps offers the best and most theoretically informed measure to aid researchers in selecting competing models. The BIC statistic is derived from the Bayesian posteriori test theory (Raftery, 1986, 1996) and can be calculated from the following:

$$BIC = L^2 - df * \log N, \qquad (2.43)$$

where L^2 is the log-likelihood goodness-of-fit test statistic, df is the degrees of freedom of a model, and N is the sample size. One major advantage of the BIC statistic is that it can be used to compare models that are either nested or nonnested.[7] Generally speaking, when choosing between various competing models, we choose the one that has the most negative value in the BIC statistic. At the same time, both Raftery (1996) and Wong (1995) caution that a small difference in BIC (say, within 5 points) is marginal for large samples.

While the development of the BIC statistic may offer a more defensible strategy than others, it can be and has been subjected to abuse by empirical researchers as well, especially when only a limited number of models are considered. In fact, any indiscriminate use of the BIC statistic and similar model selection criteria may distort our understanding of the complex underlying association pattern (Weakliem, 1999). Simply put, choosing the least wrong model among a group of *incorrect* and poorly specified models hardly constitutes a defensible strategy. The best strategy is to formulate and compare different competing models for alternative understanding.

While it is true that sample size and strength of association can affect the goodness-of-fit and the likelihood-ratio test statistics significantly when the specified models are incorrect, one must also recognize that their influence on the "true" model and its overparameterized counterparts is small and negligible (Wong, 2003a). In fact, for properly specified models, the goodness-of-fit statistics in log-linear models (association models included) are *independent* of sample sizes (Wong, 2003a). In other words, for properly specified models, nominal test statistics and nested chi-square tests offer reliable tools in model selection.

The above discussion illustrates two different schools of thought on model selection: the Bayesian approach and the classical statistical approach. Instead of construing the two approaches as opposing each other or there being trade-offs between the two, it is perhaps more meaningful to regard the two concerns as complementary. As demonstrated from various examples included in the present monograph, it is possible that both approaches can arrive at the same conclusion if we exercise great care in entertaining various competing formulations of underlying association patterns.

Asymptotic, Jackknife, and Bootstrap Standard Errors

Although the Newton unidimensional method has been the preferred method for estimating RC-type association models because of its speed and ease of use, it nonetheless has one undesirable outcome: The procedure does not produce asymptotic standard error for each log-multiplicative component as a by-product. The alternating scoring algorithm method proposed by Gilula and Haberman (1986), based on the Newton-Raphson procedure, circumvents the problem by obtaining simultaneously both parameter estimates and their corresponding asymptotic standard errors. Because the proposed procedure is computationally intensive and does not always yield converged estimates, Haberman (1995) later proposes a modified algorithm that effectively circumvents the above problems. Unfortunately, its companion Fortran program, DASSOC, is limited to estimate $RC(M)$ model for two-way tables and, therefore, has not been widely adopted by empirical researchers.

On the other hand, Henry (1981), Clogg, Shockey, and Eliason (1990), and Clogg and Shihadeh (1994) have proposed the use of the jackknife method to yield the appropriate standard errors. Both Henry (1981) and Clogg et al. (1990) indicate that the jackknife estimate of the variance behaves rather well as compared with other large-sample methods such as the delta method or the inverse of the estimated information matrix. The jackknife procedure can be easily implemented in the analysis of cross-classification tables. Suppose ξ denotes the parameter of interest in the population and $\hat{\xi}$ denotes the corresponding maximum likelihood estimator in the sample. Let us further denote $\hat{\xi}_h$ as the value obtained from the sample when the hth observation has been deleted. The population parameter ξ can be estimated either from the usual maximum likelihood estimators based on all n observations or as the mean of the n replicates. The latter can be denoted as follows:

$$\hat{\xi}^* = \sum_{h=1}^{n} \hat{\xi}_h / n. \qquad (2.44)$$

By construction, $\hat{\xi}^*$ will always be equivalent to $\hat{\xi}$. The sampling variance of $\hat{\xi}^*$ or $\hat{\xi}$ can now be approximated as the sum of the squares,

$$s^2(\hat{\xi}) = \sum_{h=1}^{n} (\hat{\xi}_h - \hat{\xi}^*)^2. \qquad (2.45)$$

For contingency tables, the jackknife estimator is simply a weighted average:

$$\hat{\xi}^* = \sum_{i=1}^{I} \sum_{j=1}^{J} f_{ij} \hat{\xi}_{ij} / n, \qquad (2.46)$$

with the variance as the weighted sum of squares:

$$s^2(\hat{\xi}) = \sum_{i=1}^{I} \sum_{j=1}^{J} f_{ij}(\hat{\xi}_{ij} - \hat{\xi}^*)^2. \qquad (2.47)$$

Of course, the square root of the above quantity is the jackknife standard error of interest.[8]

The number of jackknife replicates depends only on the number of cells (e.g., $I \times J$ for two-way tables or $I \times J \times K$ for three-way tables), not the total number of observations, n. As noted by Clogg and Shihadeh (1994, p. 37), the jackknife method is useful "not only to calculate sampling variances (or variance-covariance matrix) for quantities that are otherwise difficult to analyze, but also for the analysis of influence of observations." Unfortunately, the latter property has not been extensively investigated by empirical researchers.

Although the jackknife standard errors are simple to obtain, there has not been any systematic effort to evaluate its performance in various types of association models relative to the asymptotic and bootstrap standard errors. Based on the limited number of illustrative examples in the monograph and from the author's personal experiences, the jackknife standard errors from various association models are generally close but slightly larger than the asymptotic standard errors obtained directly from GLIM or R.[9]

The performance of the bootstrap method, on the other hand, is highly comparable to the jackknife method, provided that the number of bootstrap replicates is relatively large, say, 10,000 or more. Generally speaking, bootstrapping is a nonparametric, computer-intensive approach to statistical inference (Efron, 1981; Efron & Tibshirani, 1993; Mooney & Duval, 1993). Through the use of random sampling with replacement from the original data set, the bootstrap method constructs the approximate sampling distribution of the statistic of interest. We compared the bootstrap standard errors from more than 50,000 replicates with the asymptotic standard errors for a number of models and found them to be highly comparable (results not shown but can be obtained from the author). Similar to the case of the jackknife standard errors, the bootstrap standard errors are found to be slightly larger than the asymptotic standard errors as well, particularly for the intrinsic association (ϕ) parameters, though the differences in the estimated row and column scores parameters (μ_i and v_j) are much smaller. On the other hand, for some highly parameterized association models, say models with linear trend constraint, all three types of standard errors can be virtually identical.

Generally speaking, it is preferable to report the asymptotic standard errors when they can be obtained directly from statistical software such as R.

However, when these standard errors cannot be obtained directly as in the case of multidimensional $RC(M)$ association models with cross-dimensional constraints, the jackknife or bootstrap standard errors should be used instead. Since the jackknife method is simple to implement and requires relatively little computing time, it is slightly more preferred. In the present monograph, all reported standard errors are based on the asymptotic standard errors, whenever possible. However, when they are not readily available, the bootstrap standard errors based on 50,000 replicates are reported instead, but readers are strongly encouraged to use the jackknife method to compare with the results presented here.

The Problem of Zero and Sparse Cells

One recurrent problem in the analysis of cross-classification tables is the occurrence of zero or sparse cells, and researchers often worry that their existence may systematically distort the underlying association pattern. There are two types of zero cells: structural zeros and sampling zeros. Structural zeros exist for cells with expected value of 0; that is, no observation *could* fall into those cells. Examples of the former include the diagonal cells in the import and export trading table or an incomplete table. Sampling zeros, on the other hand, exist when the observed cell count in the table is zero but its expected value is not. Sampling variation accounts for the existence of observed zeros. The solution for structural zeros is simple: One can simply block out those problematic cells and proceed with weighted data analysis, and the degrees of freedom of each model should be adjusted accordingly. The solution of sampling zeros, on the other hand, needs careful scrutiny since some interaction parameters may become undefined. One popular strategy is the addition of a small constant (say, 0.50 or 0.10) to each cell (Bishop et al., 1975; Goodman, 1972). Clogg, Rubin, Schenker, Schultz, and Weidman (1991) offer a slightly more complex method to obtain interaction parameters for all cells. One consequence of such practice is the inflation of sample size that can become significant under sparse data situations. Paradoxically, this correction procedure becomes problematic precisely in situations when it has most often been applied (Clogg & Shihadeh, 1994, p. 17).

On the other hand, the existence of zero cells does not seem to pose any major problem for association models. This observation has been confirmed from the sensitivity analysis conducted by the author here and elsewhere (Wong, 2001). Relative to the unadjusted model, the addition of any small constant to all cells does not affect the parameter estimates in any meaningful manner. One plausible explanation for such stability in parameter estimates is that these models are highly parametric, and the derived association parameters are based on a group of cells in a particular row and/or column, not individual cells. Of course, the matter may still be problematic for row and/or column marginals that are close to zero.

Example 2.1: One-Dimensional Association Models

The first example, Table 2.3A, involves a simple cross-classification between individual political orientation (POLVIEWS) and their attitude toward gender division of labor (FEFAM). The political view variable is a self-identified political orientation measure with seven response categories: strongly liberal, liberal, slightly liberal, moderate, slightly conservative, conservative, and strongly conservative. The gender orientation variable asks the extent to which respondents agree with the statement that "it is better for men to work and women to tend home." Four categories are available: strongly disagree, disagree, agree, and strongly agree. The raw counts are derived from 1998 and 2000 *General Social Survey*, with a total of 3,439 respondents. Our interest here is how attitude toward gender division of labor is related to individual political orientation. For example, can we state that the more politically conservative (or liberal) a person is, the more likely that she or he would agree (or disagree) with the "separate spheres" ideology?

Table 2.4 reports a series of association models to understand the relationship between the two variables. While the null association or independence model (Line 1) does not provide satisfactory results, all the other models reported in Table 2.4 provide dramatic improvement as indicated from the goodness-of-fit and *BIC* statistics. For example, by using just 1 *df*, the uniform association model (Line 2) captures slightly more than 90% of the association between POLVIEWS and FEFAM and the L^2/df ratio is only slightly greater than 1.[10] Since the log-linear row effects (R) and column effects (C) models (Lines 3 and 4, respectively) both offer satisfactory results, it is not surprising to find that their more complicated counterparts ($R+C$ and RC, i.e., Lines 5 and 6, respectively) would provide acceptable results as well.

Of all the models presented here, three models (C, $R+C$, and RC) can be considered as acceptable, and the choice between them may be difficult. Given that they all offer reasonable understanding of the relationship involved, using the Occam razor principle, it appears that the uniform association model (U) and the column effects model (C) should both be preferred. On the other hand, the *BIC* statistic clearly favors the simplest model (U) rather than the slightly more complicated C model. Entries in Panels B

Table 2.3 Two-Way Cross-Classification Examples

A. Political views and attitude toward female labor ($N = 3,439$)

	FEFAM			
POLVIEWS	**Strongly Disagree**	**Disagree**	**Agree**	**Strongly Agree**
Strongly liberal	39	50	18	4
Liberal	140	178	85	23
Slightly liberal	108	195	97	23
Moderate	238	598	363	111
Slightly conservative	78	250	150	55
Conservative	50	200	208	74
Strongly conservative	8	29	46	21

POLVIEWS: Think of self as liberal or conservative

FEFAM: Better for men to work and women to tend home

B. Education and occupational attainment among women ($N = 3,858$)

	OCC				
EDUC	**Upper Nonmanual**	**Lower Nonmanual**	**Upper Manual**	**Lower Manual**	**Farm**
College+	518	95	6	35	5
Junior college	81	67	4	49	2
High school	452	1,003	67	630	5
<High school	71	157	37	562	12

EDUC: Educational attainment

OCC: Occupational attainment

SOURCE: Data for Panels A and B are derived from 1998 to 2000 and 1985 to 1990 *General Social Survey*, respectively.

Table 2.4 Analysis of Association in Table 2.3A (POLVIEWS × FEFAM)

A. Association models applied to Table 2.3A

Model Description	df	L^2	BIC	Δ	p
1. O	18	211.70	65.12	8.09	.000
2. U	17	20.12	−118.31	2.77	.268
3. R	12	15.91	−81.81	2.47	.196
4. C	15	14.24	−107.91	2.32	.508
5. R+C	10	7.68	−73.75	1.77	.660
6. RC	10	8.07	−73.36	1.77	.622

B. Components of association in Table 2.4A

Components	Models Used	df	Likelihood-Ratio Chi-Square
General effect	(1) − (2)	18 − 17 = 1	191.58
Row and column effects	(2) − (6)	17 − 10 = 7	12.05
Other effects	(6)	10	8.07
Total effects	(1)	18	211.70

C. Components of row and column effects on association in Table 2.4A

Components	Models Used	df	Likelihood-Ratio Chi-Square
Column effects	(2) − (4)	17 − 15 = 2	5.88
Row effects	(4) − (6)	15 − 10 = 5	6.17
Row and column effects	(2) − (6)	17 − 10 = 7	12.05

NOTE: *BIC* represents the Bayesian information criterion, where $BIC = L^2 - df * \ln N$ and Δ represents the index of dissimilarity.

and C provide two different ways to decompose the goodness-of-fit statistics from different models to aid our understanding of the relative contribution of various components. For example, entries from Panel B confirm that while the single uniform association parameter captures a substantial proportion of the variation in the goodness-of-fit statistics (slightly more than 90%), the 5.7% contribution from row and column effects parameters should not be dismissed offhandedly ($p < .10$). Entries in Panel C examine the relative contributions of row effects, column effects, and row and column effects. Since the difference in the goodness-of-fit statistics between $R+C$ and RC is negligible, only minor differences are expected from either calculation. For illustrative purposes, we use the RC model as the baseline for comparison. Everything else being equal, the decomposition strategy indicates that with only 2 df, the column effects parameters alone capture about 29% (=5.88/20.12) of the total effects, whereas the row effects parameters account for an additional 31% (=6.17/20.12). The latter, however, uses five additional parameters. In other words, of all association models presented in Table 2.4, the C model appears to be the preferred final model.

Table 2.5 provides not only the parameter estimates from different association models estimated in Table 2.4 but also the expected adjacent log-odds ratios to aid our interpretation of the parameters and the underlying structure of odds ratios implied from those parameters. For example, the uniform association is estimated to be 0.202, and the expected adjacent log-odds ratios all have the same values as well (see Equation 2.8). For the row effects model (R), the estimated row effects parameters will set up the structure of the expected adjacent log-odds ratios. Two sets of parameters are reported here, the first one adopts the normalization that $\sum_i \tau_i^A = 0$, whereas the second one imposes that $\tau_1^A = 0$, but they both yield identical predicted log-odds ratios. For example, entries in the first row of the predicted adjacent log-odds ratios compare row 2 versus row 1 and they all have the same value 0.154 (=[−0.405] − [−0.559]) and do not differ by the columns under consideration. Likewise, when comparing row 7 with row 6, all entries will have the value of 0.253 (=0.672 − 0.419) (see Equation 2.10). By the same token, one can use the column effects parameters from the C model to understand why the expected adjacent log-odds ratios when comparing different adjacent columns all have the same values and they do not differ by the adjacent rows under consideration (see Equation 2.12).

As for the $R+C$ and RC models, the calculation of the expected adjacent log-odds ratios will be slightly more complicated as one would need to consider both row effects and column effects simultaneously.[11] For example, under the $R+C$ model, when considering cells in the first and second rows

(Text continues on page 40)

Table 2.5 Relationship Between Expected Adjacent Log-Odds Ratios and Estimated Association Parameters for Table 2.3A (POLVIEWS × FEFAM)

Association Model		Expected Adjacent Log-Odds Ratios			
		2:1	3:2	4:3	
A. U	2:1	0.202	0.202	0.202	
	3:2	0.202	0.202	0.202	
	4:3	0.202	0.202	0.202	
	5:4	0.202	0.202	0.202	
	6:5	0.202	0.202	0.202	
	7:6	0.202	0.202	0.202	
Estimated parameter					
Uniform association		0.202			
		(0.015)			
B. R	2:1	0.154	0.154	0.154	
	3:2	0.157	0.157	0.157	
	4:3	0.257	0.257	0.257	

(Continued)

Table 2.5 (Continued)

Association Model	Expected Adjacent Log-Odds Ratios						
	2:1			**3:2**			**4:3**
	1	2	3	4	5	6	7
5:4	0.104			0.104			0.104
6:5	0.307			0.307			0.307
7:6	0.253			0.253			0.253
Estimated parameters							
Row effects	−0.559 (0.108)	−0.405 (0.059)	−0.248 (0.058)	0.009 (0.040)	0.112 (0.051)	0.419 (0.051)	0.672 (0.100)
Alternative	0.000 —	0.154 (0.136)	0.310 (0.136)	0.568 (0.127)	0.671 (0.133)	0.978 (0.133)	1.231 (0.168)
C. C 2:1	0.257			0.203			0.112
3:2	0.257			0.203			0.112
4:3	0.257			0.203			0.112
5:4	0.257			0.203			0.112
6:5	0.257			0.203			0.112
7:6	0.257			0.203			0.112

Association Model	Expected Adjacent Log-Odds Ratios			
	2:1		**3:2**	**4:3**
	1	2	3	4
Estimated parameters				
Column effects	−0.322	−0.065	0.137	0.250
	(0.026)	(0.020)	(0.023)	(0.034)
Alternative normalization	0.000	0.257	0.459	0.571
	—	(0.035)	(0.039)	(0.053)
D. *R+C* 2:1	0.194		0.119	0.016
3:2	0.206		0.131	0.028
4:3	0.317		0.242	0.139
5:4	0.179		0.104	0.001
6:5	0.407		0.333	0.230
7:6	0.374		0.299	0.196

(Continued)

Table 2.5 (Continued)

Association Model	Expected Adjacent Log-Odds Ratios						
	2:1		3:2			4:3	
	1	2	3	4	5	6	7
Estimated parameters							
Row effects	0.000	−0.086	−0.159	−0.122	−0.222	−0.095	0.000
	—	(0.114)	(0.106)	(0.092)	(0.105)	(0.115)	—
Column effects	0.000	0.279	0.484	0.586			
	—	(0.044)	(0.062)	(0.087)			
Alternative normalization							
Row effects	0.000	−0.086	−0.159	−0.122	−0.222	−0.095	0.000
	—	(0.114)	(0.106)	(0.092)	(0.105)	(0.115)	—
Column effects	0.000	0.084	0.094	0.000			
	—	(0.032)	(0.039)	—			
Uniform association	0.195						
	(0.029)						

Association Model — **Expected Adjacent Log-Odds Ratios**

E. RC	2:1	3:2	4:3
2:1	0.153	0.119	0.056
3:2	0.204	0.159	0.075
4:3	0.314	0.245	0.116
5:4	0.135	0.105	0.050
6:5	0.442	0.345	0.163
7:6	0.378	0.295	0.140

Estimated parameters

	1	2	3	4	5	6	7
Row scores, μ_i	−0.482	−0.376	−0.234	−0.016	0.078	0.384	0.646
	(0.068)	(0.050)	(0.048)	(0.037)	(0.049)	(0.060)	(0.061)
Column scores, ν_j	−0.748	−0.141	0.332	0.557			
	(0.027)	(0.047)	(0.052)	(0.048)			
Intrinsic association, ϕ	2.373						
	(0.238)						

NOTE: Values in parentheses are the asymptotic standard errors. See text for details.

and first and second columns, the adjacent log-odds ratio equals to 0.194 ($= -0.086 + 0.279$). Likewise, for cells in rows 6 and 7 and columns 3 and 4, the adjacent log-odds ratio equals to 0.196 ($=[0 - -0.095] + [0.586 - 0.484]$). The calculation of adjacent log-odds ratios under an alternative normalization will be an identical, albeit slightly complex, calculation (see Equation 2.15). On the other hand, under the *RC* model, the adjacent log-odds ratios equal to 0.153, that is, $\{2.373 * [-0.376 - -0.482] * [-0.141 - -0.748]\}$ and 0.140 ($=\{2.373 * [0.646 - 0.384] * [0.557 - 0.332]\}$, respectively (see Equation 2.19). Finally, readers are encouraged to apply the formulae to calculate the expected adjacent log-odds ratios for other entries as well.

Our understanding of the relationship between political orientation and attitude toward gender division of labor remains largely similar, whether one adopts the *U*, *C*, *R+C*, or *RC* models as the final preferred model. This is because all estimated scores are monotonic, and the only difference between them is whether they are equidistant. Therefore, the general statement that the more politically conservative (or liberal) a person is, the more likely that she or he would agree (or disagree) with the "separate spheres" ideology is largely valid.

Example 2.2: Two-Dimensional Association Models

The second sample (Table 2.3, Panel B) involves an analysis of the relationship between educational and occupational attainment among American women. The tabulation is derived from the 1985–1990 cumulative *General Social Survey* and has been analyzed by Wong (1995, 2001). Educational attainment is measured by educational qualification (highest educational degree attained) rather than years of education to approximate the relationship between educational credentials and labor market outcomes. It has four outcomes: college and more, less than college, high school, and less than high school. Occupational attainment is classified by skill levels and industrial sectors and has five categories: upper nonmanual, lower nonmanual, upper manual, lower manual, and farm. Unlike Wong (1995, 2001), the analysis here combines both white and African Americans together, and the table has a total of 3,858 respondents. Earlier results (Clogg & Shihadeh, 1994; Wong, 2001) suggest that the two-dimensional association models are adequate to understand the complex relationship involved.

In addition to the independence model, Table 2.6 reports a series of one-dimensional (Lines 2 to 6) and two-dimensional association models (Lines 7 to 11) to account for the association between education and occupation. While there is a dramatic improvement in goodness-of-fit statistics

Table 2.6 Analysis of Association in Table 2.3B (EDUC × OCC)

A. Association models applied to Table 2.4B

Model Description	df	L^2	BIC	Δ	p
1. O	12	1373.18	1274.08	23.86	.000
2. U	11	244.02	153.18	8.54	.000
3. R	9	205.97	131.65	7.38	.000
4. C	8	155.37	89.31	7.47	.000
5. R+C	6	91.61	42.06	4.63	.000
6. RC	6	125.06	75.51	6.44	.000
7. U+RC	5	17.60	−23.69	1.52	.004
8. R+RC	4	6.94	−26.10	0.83	.139
9. C+RC	3	11.41	−13.37	1.01	.010
10. R+C+RC	2	0.28	−16.24	0.01	.870
11. RC(2)	2	0.60	−15.92	0.09	.741

(Continued)

41

Table 2.6 (Continued)

B. Components of association in Table 2.4B

Components	Models Used	df	Likelihood-Ratio Chi-Square
First dimension	(1) – (6)	12 – 6 = 6	1248.12
Second dimension	(6) – (11)	6 – 2 = 4	124.46
Higher dimension	(11)	2	0.60
Total effects	(1)	12	1373.18

C. Components of row and column effects on association in Table 2.4B

Components	Models Used	df	Likelihood-Ratio Chi-Square
General effect in U	(1) – (2)	12 – 11 = 1	1129.16
Row and column effects in RC	(2) – (6)	11 – 6 = 5	118.96
Additive row and column effects in $R+C+RC$	(6) – (10)	6 – 2 = 4	124.78
Other effects	(10)	2	0.28
Total effects	(1)	12	1373.18

NOTE: *BIC* represents the Bayesian information criterion, where $BIC = L^2 - df * \ln N$ and Δ represents the index of dissimilarity.

as compared with the simple independence model (*O*), none of the one-dimensional association models (*U*, *R*, *C*, *R+C*, and *RC*) yield satisfactory results. The *p* value of each model is statistically significant at the .001 level. Also, a substantial proportion of individuals are still misclassified, as noted from the index of dissimilarity.

In contrast, by increasing model complexity from one-dimensional to two-dimensional association models, we find at least two models that can provide adequate account of the association between education and occupation. For instance, the addition of a single uniform association parameter to the *RC* model (Line 7, *U+RC*) results in dramatic improvement in the goodness-of-fit statistic over the *RC* model in Line 6 (1 *df*, $\Delta L^2 = 107.46$), and the *BIC* statistic now becomes negative and is preferred over other simpler models reported so far. Nonetheless, the slightly more complicated model is still not unsatisfactory if we use the conventional chi-square test as the guideline in model selection (5 *df*, $L^2 = 17.40$). The addition of row effects parameters and column effects parameters (Lines 8 and 9, respectively) fails to provide satisfactory results either, though the overall fit of the latter model is marginal. On the other hand, the addition of both row and column effects (log-linear or log-multiplicative) to the *RC* model (i.e., Lines 10 and 11) leads to satisfactory results. The *p* values for both models (*R+C+RC* and *RC(2)*) are highly acceptable, and only a tiny proportion of individuals in the sample has been misclassified.[12]

Using the *O*, *RC*, and *RC(2)* models for comparison, Panel B of Table 2.6 presents the proportion of L^2 that can be accounted by one-, two-, and higher-dimensional association models. The decomposition indicates clearly that while the first dimension accounts for close to 90% of the L^2, the second dimension contributes an additional 9%, whereas the contribution from other higher dimensions is negligible. Panel C offers another decomposition strategy involving the *R+C+RC* model. It yields basically the same conclusion that additional dimension of row and column effects, whether log-linear or log-multiplicative, is important to aid our understanding of the complex relationship involving women's educational qualification and occupational attainment.

Similar to Table 2.5, Table 2.7 not only reports the parameter estimates and their asymptotic standard errors from different two-dimensional association models but also calculates the expected adjacent log-odds ratios so that we can understand how the imposed structure of odds ratios look like from each model. For example, given the parameter estimates from the *U+RC* model, one can calculate that the expected adjacent log-odds ratio for rows 1 and 2 and columns 1 and 2 equals to 1.132, that is, {0.552 + 3.435 * [0.147 − 0.722] * [−0.451 − −0.158]}. Similarly, the expected adjacent log-odds ratio for rows 3 and 4 and columns 4 and 5 equals to 1.790, that is, {0.552 + 3.435 * [−0.236 − −0.633] * [0.855 − −0.053]}. The exact interpretation of parameters is more complicated because the model is two-dimensional. Given that the first dimension is captured by a single uniform association parameter, it assumes that the relationship between educational qualification and occupational attainment is largely uniform, linear, and equidistant. The second dimension, on the other hand, captures departures from such linearity. For example, combining information from both row and column scores parameters, the second dimension indicates that it is far more likely for women with high school and less than high school education to end up in lower nonmanual positions. Given that the $U+RC$ model is not the preferred model, it means that although the vertical socioeconomic image of educational qualification and occupational attainment is basically valid (i.e., the higher the education, the higher the occupational attainment), the equidistant assumption is possibly too strong to hold in reality.

By the same token, one can calculate the expected adjacent log-odds ratios for other two-dimensional association models easily from the estimated parameters. For simplicity's sake, the remaining discussion focuses only on the $RC(2)$ model.[13] Using Equation 2.29 or 2.34, the calculation of the expected adjacent log-odds ratios is straightforward. For instance, the odds ratio involving rows 1 and 2 and columns 1 and 2 would be 1.481, that is, {2.601 * [−0.089 − −0.744] * [−0.020 − −0.765] + 1.522 * [−0.061 − 0.276] * [−0.549 − −0.137]}, whereas the odds ratio involving rows 3 and 4 and columns 4 and 5 would be 0.966 or {2.601 * [0.632 − 0.200] * [−0.088 − 0.550] + 1.522 * [0.562 − −0.077] * [0.816 − −0.010]}.

Based on the rank ordering of row and column categories, it is clear that the first dimension is more or less socioeconomic in nature, that is, better educated women tend to be more likely to end up in upper-nonmanual positions. However, since the estimated column scores for farm is negative (−0.088), it means that everything else being equal, American women (blacks and whites) are much more likely to end up in farming occupations than in upper- or lower-manual occupations. The impermeability boundary between nonmanual and farm is actually much weaker than those involving the manual sector. Also, given that the estimated distances between row and column categories in the first dimension are not equidistant, we can now understand why the $U+RC$ model fails to provide an adequate understanding than its more complicated counterparts, the $R+C+RC$ or $RC(2)$ models.

Table 2.7 Relationship Between Expected Odds-Ratios and Estimated Association Parameters for Table 2.3B (EDUC × OCC)

Association Model	Expected Adjacent Log-Odds Ratios			
	2:1	**3:2**	**4:3**	**5:4**
A. $U+RC$				
2:1	1.132	0.043	0.274	−1.244
3:2	1.337	−0.137	0.175	−1.881
4:3	0.153	0.903	0.744	1.790

Estimated Parameters	1	2	3	4	5
Row scores, μ_i	0.722	0.147	−0.633	−0.236	
	(0.049)	(0.095)	(0.046)	(0.066)	
Column scores, ν_j	−0.158	−0.451	−0.194	−0.053	0.855
	(0.033)	(0.039)	(0.058)	(0.041)	(0.013)

Uniform association		Intrinsic association (ϕ)
0.552		3.436
(0.034)		(0.619)

B. $R+RC$	**2:1**	**3:2**	**4:3**	**5:4**
2:1	1.488	0.175	0.498	−1.465
3:2	1.019	−0.249	0.063	−1.832
4:3	0.156	0.916	0.729	1.864

Table 2.7 (Continued)

Association Model		Expected Adjacent Log-Odds Ratios			
		2:1	3:2	4:3	5:4
Estimated parameters	1	2	3	4	5
Row effects	0.000	-0.432	-1.267	0.000	
	—	(0.221)	(0.193)	—	
Row scores, μ_i	-0.760	-0.061	0.613	0.209	
	(0.042)	(0.097)	(0.049)	(0.068)	
Column scores, ν_j	-0.806	-0.057	0.180	0.543	0.140
	(0.017)	(0.045)	(0.063)	(0.046)	(0.112)
Intrinsic association, ϕ	3.670				
	(0.547)				
C. *C+RC*					
2:1		1.171	0.109	0.158	-1.247
3:2		1.296	-0.013	0.125	-1.477
4:3		-0.009	1.262	0.471	0.941

46

Association Model		Expected Adjacent Log-Odds Ratios			
		2:1	**3:2**	**4:3**	**5:4**
Estimated parameters	1	2	3	4	5
Column effects	0.000	1.263	1.282	1.416	0.000
	—	(0.170)	(0.209)	(0.091)	—
Row scores, μ_i	−0.333	−0.251	−0.281	0.865	
	(0.133)	(0.089)	(0.144)	(0.007)	
Column scores, ν_j	−0.117	−0.559	−0.127	−0.009	0.811
	(0.055)	(0.077)	(0.093)	(0.064)	(0.041)
Intrinsic association, ϕ	2.509				
	(0.473)				
D. R+C+RC					
2:1		1.503	0.260	0.414	−1.280
3:2		0.989	−0.126	−0.021	−1.603
4:3		−0.003	1.270	0.470	0.982

(Continued)

Table 2.7 (Continued)

Association Model			Expected Adjacent Log-Odds Ratios			
Estimated parameters	1	2:1	3:2	4:3	5:4	
		2	3	4	5	
Row effects	0.000	-0.290	-0.955	0.000	0.000	
	—	(0.184)	(0.217)	—	—	
Column effects	0.000	0.000	0.397	0.327	0.000	
	—	—	(0.167)	(0.161)	—	
Row scores, μ_i	-0.766	-0.055	0.601	0.221		
	(0.037)	(0.090)	(0.047)	(0.068)		
Column scores, ν_j	-0.838	0.044	0.119	0.501	0.174	
	(0.025)	(0.089)	(0.092)	(0.066)	(0.126)	
Intrinsic association, ϕ	2.857					
	(0.534)					
E. RC(2)	2:1	1.481	0.362	0.331	-1.510	
	3:2	1.008	-0.211	0.051	-1.378	
	4:3	-0.002	1.259	0.480	0.966	

48

Association Model		Expected Adjacent Log-Odds Ratios			
		2:1	3:2	4:3	5:4
Estimated parameters	1	2	3	4	5
First dimension					
Row scores, μ_{i1}	-0.744	-0.089	0.200	0.632	
	(0.041)	(0.042)	(0.101)	(0.074)	
Column scores, ν_{j1}	-0.765	-0.020	0.323	0.550	-0.088
	(0.052)	(0.077)	(0.101)	(0.060)	(0.179)
Intrinsic association, ϕ_1	2.601				
	(0.151)				
Second dimension					
Row scores, μ_{i2}	0.276	-0.061	-0.777	0.562	
	(0.121)	(0.158)	(0.063)	(0.086)	
Column scores, ν_{j2}	-0.137	-0.549	-0.120	-0.010	0.816
	(0.160)	(0.075)	(0.092)	(0.120)	(0.036)
Intrinsic association, ϕ_2	1.522				
	(0.325)				

NOTE: Values in parentheses are the asymptotic standard errors. See text for details.

To fully interpret the meaning of the reported parameters in the $RC(2)$ model, their estimated row and column scores from both dimensions are plotted in Figures 2.2 and 2.3, respectively, using the symmetrical normalization (Equation 2.38). In both cases, one can see that the first dimensional row scores and column scores have clear rank orderings, except the Farm category. On the other hand, the rank orderings of row and column scores in the second dimension are not straightforward. As indicated earlier, one can interpret the row and column scores in the second dimension as departures from the first dimension, that is, either over- or undercorrections from the vertical image depicted in the first dimension. To a large extent, they represent specific channels and barriers that many American women have to face in the labor market that deviate from the meritocratic image as depicted by the first dimension. The pattern displayed lends some support to the feminist's claim of the gendered division of labor in American society.

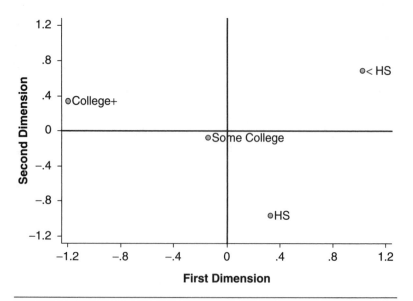

Figure 2.2   Estimated Education Scores From Model 11 in Table 2.6

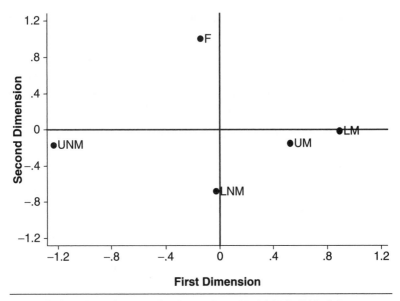

Figure 2.3 Estimated Occupation Scores From Model 11 in Table 2.6

Notes

1. There is some arbitrariness in identifying the sign for ϕ since it is the product $(\phi\mu_i\nu_j)$ that is uniquely identified. To reduce confusion and facilitate the calculation of the bootstrap standard errors, the intrinsic association (ϕ) is defined as a positive value but the signs of the estimated row (μ_i) and column scores (ν_j) will be changed accordingly.

2. For instance, in addition to the *RC* component, Kateri, Ahmad, and Papaioannou (1998) formulate a series of multidimensional *U*, *R*, *C*, and *R+C* parameters by using orthogonal polynomials.

3. de Rooij and Heiser (2005) and de Rooij (2008) caution against the (mis) interpretation of those symmetrically normalized scores as "distances" in pure mathematical sense because they represent inner-product distances rather than between scores distances.

4. CDAS is available from Scott Eliason via personal request. l_{EM} can be downloaded from www.uvt.nl/faculteiten/fsw/organisatie/departementen/mto/software2. html. R is a free statistical software that can be downloaded from www.r-project .org/. CDAS can run only under DOS-emulated environment, l_{EM} under both DOS and Windows environment, and R is the only one that can operate in UNIX, Windows, and MacOS environment. Also, readers may be interested in Latent Gold

(Vermunt & Magidson, 2005), a popular commercial statistical software to estimate latent class and other finite mixture models that treats log-multiplicative association models as latent variable models.

5. Readers can consult with the computing examples illustrated in Powers and Xie (2008) at the following Web site: https://webspace.utexas.edu/dpowers/www/. Also, the DASSOC program written by Haberman (1995) can be downloaded from the STATLIB Web site directly (http://lib.stat.cmu.edu/general/).

6. Like 1_{EM}, the current *gnm* module does not have the capability to impose cross-dimensional constraints, and we cannot obtain asymptotic standard errors for log-multiplicative parameters. Therefore, for models that require cross-dimensional constraint(s), the bootstrap or jackknife standard errors are used instead.

7. If we can express models M_1, M_2, and M_3 such that $M_1 \subset M_2 \subset M_3$, then the three models are nested. However, if there are only partial overlaps between them and they cannot be subsumed under each other, then the three models are nonnested. The difference in test statistics between nonnested models does not have proper chi-square interpretation and would involve nonnested chi-square tests instead (Weakliem, 1992).

8. Henry (1981) reports a slightly different formula for the jackknifed standard errors. His calculation of the variance includes a small correction factor $(N-1)/N$, where N is the sample size. For most practical purposes, the difference should be small when N is large.

9. For example, the jackknife standard errors for a number of log-linear association models (e.g., U, R, C, and $R+C$) turn out to be slightly larger than their asymptotic counterparts obtained from GLIM and R.

10. If the model is true, the L^2/df ratio should be very close to 1. For some empirical practitioners, however, a ratio of less than 2 may be considered as satisfactory.

11. If we adopt the normalization that $\sum \tau_j^B = 0$ instead, then the column effects parameters are −0.337, −0.058, 0.147, and 0.249 and their corresponding asymptotic standard errors are 0.044, 0.024, 0.026, and 0.048, respectively.

12. One may consider that both specifications overfit the data. However, as illustrated in Wong (2001), the $RC(2)$ model fits well for white and black men and women as well. More important, when the same model has been applied to all four groups simultaneously as the conditional $RC(2)$ model, it is possible to generate even simpler models with further restrictions that would yield highly interpretable results. This procedure will be discussed more extensively in Chapter 4.

13. The asymptotic standard errors for the $RC(2)$ model are obtained directly from the DASSOC program (Haberman, 1995). This is because the current *gnm* module in R does not impose proper cross-dimensional constraints and, therefore, cannot be used directly. We also calculate the bootstrap standard errors (with more than 50,000 replicates) and find them to be highly comparable with the asymptotic standard errors, albeit that their values are invariably larger. The same applies to the jackknife standard errors as well.

CHAPTER 3. PARTIAL ASSOCIATION MODELS FOR THREE-WAY TABLES

The types of association models described in Chapter 2 can be extended easily to the analysis of three-way or multiway cross-classification tables (Agresti, 1983; Agresti & Kezouh, 1983; Becker, 1989a; Becker & Clogg, 1989; Pannekoek, 1985). The only complication is to decide which set of parameters (two-way, three-way, and/or higher-order interaction parameters) should be decomposed into association parameters. In this chapter, we will introduce simple models that decompose two-way interaction parameters when there is no need to incorporate three-way or higher-order parameters. In Chapter 4, the decomposition will include both two-way and three-way (or higher-order) interaction parameters. The latter is also where one can find most social science applications. In most cases, the third or layer variable is a grouping variable, such as cohort, survey year, race/ethnicity, country, and/or region. Note that the term *grouping variable* is defined in the most generic form and may actually involve two or more variables. For instance, it can be gender by ethnicity by survey year or country by cohort by survey year. Finally, although the current discussion focuses mainly on three variables, the models introduced can be generalized to analyze higher-order cross-classified tables as well.

Complete Independence (*I*) Model

Suppose we have three variables, *A*, *B*, and *C* as the row, column, and layer variable, and they each have *I*, *J*, and *K* categories, respectively. Conventional hierarchical log-linear modeling strategy that studies the relationship between them usually begins with the complete independence model, then adds various two-way interaction terms, and finally, the addition of three-way interaction parameters when none of the models with lower-order terms provide satisfactory results. Again, our discussion and analysis below does not need to explicitly specify their relationship as dependent and independent variables.

The complete independence (*I*) model postulates that there is *no* relationship between variables *A*, *B*, and *C*. Under *I*, the logarithm of the expected frequencies can be written as follows:

$$\log F_{ijk} = \lambda + \lambda_i^A + \lambda_j^B + \lambda_k^C, \tag{3.1}$$

where λ is the intercept, λ_i^A, λ_j^B, and λ_k^C are the marginal parameters, all subject to conventional normalizations similar to the ones introduced in Chapter 2 (Agresti, 2002). The model has $IJK - I - J - K + 2$ *df*.

There are two types of odds-ratios that can be examined and modeled in a three-way table (Agresti, 1983; Becker, 1989a; Becker & Clogg, 1989; Clogg, 1982a; Wong, 2001). The first one is the conditional local odds ratios for A and B given C, $\theta_{ij(k)}$, and the second one is the ratios of the conditional local odds ratios or simply the local odds ratios for A, B, and C, θ_{ijk}. The latter is simply a ratio of the conditional local odds ratios for layer $k + 1$ versus layer k (or more generally, layer k vs. layer k'). Formally, the conditional local odds ratios for A and B given C, $\theta_{ij(k)}$ can be defined as follows:

$$\theta_{ij(k)} = \frac{F_{ijk}F_{i+1,j+1,k}}{F_{i+1,jk}F_{i,j+1,k}}, \tag{3.2}$$

and the local odds ratios for A, B, and C can be defined as the ratios of the conditional local odds ratios, that is,

$$\theta_{ijk} = \frac{\theta_{ij(k+1)}}{\theta_{ij(k)}}. \tag{3.3}$$

As usual, we prefer to work with their logarithmic transformations. It is easy to verify that under I, they all have values of 1 (or 0 for their logarithms). That is,

$$\log \theta_{ij(k)} = 0,$$

$$\log \theta_{i(j)k} = 0,$$

$$\log \theta_{(i)jk} = 0,$$

and

$$\log \theta_{ijk} = \log \theta_{ij(k+1)} - \log \theta_{ij(k)} = \log \theta_{i(j+1)k} - \log \theta_{i(j)k}$$
$$= \log \theta_{(i+1)jk} - \log \theta_{(i)jk} = 0. \tag{3.4}$$

Conditional Independence (*CI*) Model

For most social science applications, it is unlikely that the complete independence model would fit the data well. The next step often involves the inclusion of two-way interaction terms (between A and B, A and C, and/or B and C) to capture the departure from complete independence. For instance, the model with only *AB* and *AC* two-way interaction terms can be written as follows:

$$\log F_{ijk} = \lambda + \lambda_i^A + \lambda_j^B + \lambda_k^C + \lambda_{ij}^{AB} + \lambda_{ik}^{AC}. \qquad (3.5)$$

Equation 3.5 is known as the conditional independence (*CI*) model because it assumes that after controlling for variable *A*, there is no relationship between variables *B* and *C*. In other words, the relationship between variables *B* and *C* is spurious after controlling for variable *A*. The model has $I(J-1)(K-1)$ *df*. Under the above formulation, it is evident that the conditional and local odds ratios have the following relationship:

$$\log \theta_{(i)jk} = 0,$$

and

$$\log \theta_{ijk} = 0. \qquad (3.6)$$

Note that $\log \theta_{i(j)k}$ and $\log \theta_{ij(k)}$ cannot be simplified under conditional independence.

Conditional Independence With Association (*CIA*) Model

If the conditional independence model fits the data well, it is possible to decompose the two interaction parameters into partial association parameters to provide even simpler interpretation (see Wong, 2001, for details). For instance, the *AB* partial association can be expressed by the $RC(M_1)$ association component, whereas the *AC* partial association can be expressed by the $RL(M_2)$ association component. The resultant model of conditional independence with association (*CIA*) can be denoted as follows:

$$\log F_{ijk} = \lambda + \lambda_i^A + \lambda_j^B + \lambda_k^C + \sum_{r=1}^{M_1} \phi_r^{AB} \mu_{ir} \nu_{jr} + \sum_{s=1}^{M_2} \phi_s^{AC} \mu_{is}^* \omega_{ks}, \qquad (3.7)$$

where M_1 and M_2 represent the dimensions needed to capture the *AB* and *AC* partial association, respectively, with $0 \le M_1 \le \min(I-1, J-1)$ and $0 \le M_2 \le \min(I-1, K-1)$. The model has $IJK - I - J - K + 2 - M_1(I + J - M_1 - 2) - M_2(I + K - M_2 - 2)$ *df*. To uniquely identify all parameters, centering, scaling, and cross-dimensional constraints are needed. When $M_1 = M_2 = 1$, Equation 3.7 can be simplified as follows:

$$\log F_{ijk} = \lambda + \lambda_i^A + \lambda_j^B + \lambda_k^C + \phi^{AB} \mu_i \nu_j + \phi^{AC} \mu_i^* \omega_k. \qquad (3.8)$$

This model has $IJK - 3I - 2J - 2K + 8$ df. It is also possible to impose consistent row score restrictions, that is, $\mu_i = \mu_i^*$ for all i to achieve an even simpler model. The restricted model would gain $I - 2$ df and has $IJK - 2I - 2J - 2K + 6$ df. Note that because the *CIA* model still assumes conditional independence, the structure of odds ratios under Equation 3.6 should therefore still hold.

Full Two-Way Interaction (*FI*) Model

When the conditional independence model is not acceptable, it becomes necessary to include all two-way interaction terms but without the incorporation of the three-way interaction parameters. The *FI* model can be written as follows:

$$\log F_{ijk} = \lambda + \lambda_i^A + \lambda_j^B + \lambda_k^C + \lambda_{ij}^{AB} + \lambda_{ik}^{AC} + \lambda_{jk}^{BC}. \tag{3.9}$$

It has $(I - 1)(J - 1)(K - 1)$ df. If variable A is a grouping variable, say cohort, survey year, or country, Equation 3.9 is sometimes also known as the constant or homogeneous association model because it postulates that the association pattern is the same across different levels of A. For instance, if variable A represents ethnicity (Asians, blacks, Hispanics, whites, and others), and variables B and C represent individual education and occupation, respectively, then the *FI* model postulates that the association between education and occupation is the same across all five ethnic groups. Under Equation 3.9, only $\log \theta_{ijk} = 0$, whereas all other conditional adjacent log-odds ratios cannot be further simplified.

Partial Association Models

Following the treatment by Agresti (1983, 1984), Agresti and Kezouh (1983), Choulakian (1996), Clogg (1982b), Pannekoek (1985), and Wong (2001), we can decompose all two-way interaction terms into partial association components. Take the simplest case for illustration, the $RC(1) + RL(1) + CL(1)$ partial association model has the following form:

$$\log F_{ijk} = \lambda + \lambda_i^A + \lambda_j^B + \lambda_k^C + \phi_1^{AB}\mu_{i1}\nu_{j1} + \phi_1^{AC}\mu_{i1}^*\eta_{k1} + \phi_1^{BC}\nu_{j1}^*\eta_{k1}^*, \tag{3.10}$$

where $\sum_{i=1}^{I} \mu_{i1} = \sum_{i=1}^{I} \mu_{i1}^* = \sum_{j=1}^{J} \nu_{j1} = \sum_{j=1}^{J} \nu_{j1}^* = \sum_{k=1}^{K} \eta_{k1} = \sum_{k=1}^{K} \eta_{k1}^* = 0$, and $\sum_{i=1}^{I} \mu_{i1}^2 = \sum_{i=1}^{I} \mu_{i1}^{*2} = \sum_{j=1}^{J} \nu_{j1}^2 = \sum_{j=1}^{J} \nu_{j1}^{*2} = \sum_{k=1}^{K} \eta_{k1}^2 = \sum_{k=1}^{K} \eta_{k1}^{*2} = 1$. That is, both centering and scaling constraints on row, column, and layer scores are needed to identify the model. Note that μ_{i1} and μ_{i1}^* are the estimated row scores for *AB* and *AC* partial association, ν_{j1} and ν_{j1}^* are the estimated column scores for AB and BC partial association, η_{k1} and η_{k1}^* are estimated layer scores for AC and BC partial association, respectively, and ϕ_1^{AB}, ϕ_1^{AC}, and ϕ_1^{BC} are the intrinsic association parameters for the AB, AC, and BC partial association, respectively. The model has $IJK - 3I - 3J - 3K + 11$ df. The above model is termed as the unrestricted $RC(1) + RL(1) + CL(1)$ partial association model (Wong, 2001) because it has only one dimension in the decomposition of each two-way interaction term, and there is no (consistent) restriction on row, column, and layer scores across different partial associations.

A more restricted model is to impose consistent score restrictions on row, column, and layer scores, that is, $\mu_{i1} = \mu_{i1}^*$, $v_{j1} = v_{j1}^*$, and $\eta_{k1} = \eta_{k1}^*$. With such restrictions, Equation 3.10 now becomes the following:

$$\log F_{ijk} = \lambda + \lambda_i^A + \lambda_j^B + \lambda_k^C + \phi_1^{AB}\mu_{i1}v_{j1} + \phi_1^{AC}\mu_{i1}\eta_{k1} + \phi_1^{BC}v_{j1}\eta_{k1}.(3.11)$$

Equation 3.11 is labeled as the restricted $RC(1) + RL(1) + CL(1)$ model with consistent score restrictions in row, column, and layer scores (Clogg, 1982a, 1982b). The model has $IJK - 2I - 2J - 2K + 5$ df. The contrast of the likelihood test statistics between Equations 3.10 and 3.11 yields a χ^2 statistic with $I + J + K - 6$ df that can be used to test whether consistent score restrictions are indeed congruent with the data. Of course, it is also plausible to impose consistent score restrictions on some but not all three variables, for instance, $\mu_{i1} = \mu_{i1}^*$ and $v_{j1} = v_{j1}^*$, but $\eta_{k1} \neq \eta_{k1}^*$.

When neither the unrestricted nor restricted $RC(1) + RL(1) + CL(1)$ model fits the data, one can increase the dimensionality in each two-way partial interaction term to underscore the complexity of relationships involved. The most general form can be labeled the $RC(M_1) + RL(M_2) + CL(M_3)$ model (Wong, 2001), where M_1, M_2, and M_3 represent the dimensionality of the AB, AC, and BC partial association, respectively; with $0 \leq M_1 \leq \min(I - 1, J - 1)$, $0 \leq M_2 \leq \min(I - 1, K - 1)$, and $0 \leq M_3 \leq \min(J - 1, K - 1)$. Without loss of generality, each set of intrinsic association parameters can be ordered from the largest to the smallest. Indeed, this model is a generalization of the $RC(M)$ association model discussed in the previous chapter (see Becker, 1989a, 1992; Goodman, 1986, 1991). It therefore follows that the $RC(M_1^*) + RL(M_2^*) + CL(M_3^*)$ model, where $M_1^* = \min(I - 1, J - 1)$, $M_2^* = \min(I - 1, J - 1)$, and $M_3^* = \min(J - 1, K - 1)$, is a saturated model for each two-way interaction term and is equivalent to the FI model of Equation 3.9. In general, the $RC(M_1) + RL(M_2) + CL(M_3)$ model can be written as follows:

$$\log F_{ijk} = \lambda + \lambda_i^A + \lambda_j^B + \lambda_k^C + \sum_{m=1}^{M_1} \phi_m^{AB}\mu_{im}v_{jm}$$
$$+ \sum_{m=1}^{M_2} \phi_m^{AC}\mu_{im}^*\eta_{km} + \sum_{m=1}^{M_3} \phi_m^{BC}v_{jm}^*\eta_{km}^*. \quad (3.12)$$

To uniquely identify the parameters, centering and scaling restrictions as well as cross-dimensional constraints on row, column, and layer score parameters within each two-way interaction are needed. The centering restrictions are as follows:

$$\sum_{i=1}^{I} \mu_{im} = \sum_{i=1}^{I} \mu_{im}^* = \sum_{j=1}^{J} v_{jm} = \sum_{j=1}^{J} v_{jm}^* = \sum_{k=1}^{K} \eta_{km} = \sum_{k=1}^{K} \eta_{km}^* = 0.$$

The scaling and cross-dimensional constraints can be written compactly as follows:

$$\sum_{i=1}^{I} \mu_{im}\mu_{im'} = \sum_{i=1}^{I} \mu_{im}^*\mu_{im'}^* = \sum_{j=1}^{J} v_{jm}v_{jm'} = \sum_{j=1}^{J} v_{jm}^*v_{jm'}^*$$
$$= \sum_{k=1}^{K} \eta_{km}\eta_{km'} = \sum_{k=1}^{K} \eta_{km}^*\eta_{km'}^* = \delta_{mm'},$$

where $\delta_{mm'}$ is the Kronecker delta with $\delta_{mm'} = 1$ if $m = m'$, 0 otherwise (note that m and m' can differ depending on the dimensionality M_1, M_2, and M_3 that is applicable to specific partial associations). The model has $IJK - I - J - K + 2 - M_1(I + J - M_1 - 2) - M_2(I + K - M_2 - 2) - M_3(J + K - M_3 - 2)$ df. It is also possible to incorporate both log-linear and log-multiplicative components to Equation 3.12 to create hybrid partial association models as discussed in the previous chapter. This alternative may be appealing to some researchers because there is no need to impose cross-dimensional constraints in certain hybrid models.

Based on Equation 3.12, the conditional local odds ratios for A and B given C, A and C given B, B and C given A, and the local odds ratios for A, B, and C under the $RC(M_1) + RL(M_2) + CL(M_3)$ model would be as follows:

$$\log\theta_{ij(k)} = \sum_{m=1}^{M_1} \phi_m^{AB} \left(\mu_{i+1,m} - \mu_{im}\right)\left(v_{j+1,m} - v_{jm}\right),$$

$$\log\theta_{i(j)k} = \sum_{m=1}^{M_2} \phi_m^{AC} \left(\mu_{i+1,m}^* - \mu_{im}^*\right)\left(\eta_{k+1,m} - \eta_{km}\right),$$

$$\log\theta_{(i)jk} = \sum_{m=1}^{M_3} \phi_m^{BC} \left(v_{j+1,m}^* - v_{jm}^*\right)\left(\eta_{k+1,m}^* - \eta_{km}^*\right),$$

and

$$\log\theta_{ijk} = 0. \tag{3.13}$$

Thus, the $RC(M_1) + RL(M_2) + CL(M_3)$ model does not postulate any three-way interaction between A, B, and C, and the conditional log-odds ratios can be written simply as a function of the product sum of row, column, and/or layer scores and their respective intrinsic association parameters across different dimensions. If $M_1 = M_2 = M_3 = M$, then the degrees of freedom for the unrestricted $RC(M) + RL(M) + CL(M)$ model can be simplified to $IJK - I - J - K + 2 - M(2I + 2J + 2K - 3M - 6)$.

Although similar consistent score restrictions can be applied easily to the $RC(M) + RL(M) + CL(M)$ model, the calculation of the exact degrees of freedom of a particular model becomes more complicated because not all cross-dimensional constraints are needed to uniquely identify all parameters. As noted by Wong (2001), only one set of cross-dimensional constraints on row, column, or layer scores will be sufficient to uniquely identify the model for restricted $RC(M) + RL(M) + CL(M)$ models with consistent score restrictions on row, column, and layer scores. For example, only one, not three, cross-dimensional constraint is needed on $\{\mu_{i1}, \mu_{i2}\}$, $\{v_{j1}, v_{j2}\}$, or $\{\eta_{k1}, \eta_{k2}\}$ for the restricted $RC(2) + RL(2) + CL(2)$ model with consistent scores. That is, either (i) $\sum_{i=1}^{I} \mu_{i1}\mu_{i2} = 0$, (ii) $\sum_{j=1}^{J} v_{j1}v_{j2} = 0$, or (iii) $\sum_{k=1}^{K} \eta_{k1}\eta_{k2} = 0$ can be imposed.

Similarly, for the restricted $RC(3) + RL(3) + CL(3)$ model, only three cross-dimensional constraints, not nine, are needed on $\{\mu_{i1}, \mu_{i2}, \mu_{i3}\}$, $\{v_{j1}, v_{j2}, v_{j3}\}$, or $\{\eta_{k1}, \eta_{k2}, \eta_{k3}\}$. That is, either (i) $\sum_{i=1}^{I} \mu_{i1}\mu_{i2} = \sum_{i=1}^{I} \mu_{i1}\mu_{i3} = \sum_{i=1}^{I} \mu_{i2}\mu_{i3} = 0$, (ii) $\sum_{j=1}^{J} v_{j1}v_{j2} = \sum_{j=1}^{J} v_{j1}v_{j3} = \sum_{j=1}^{J} v_{j2}v_{i3} = 0$, or (iii) $\sum_{k=1}^{K} \eta_{k1}\eta_{k2} = \sum_{k=1}^{K} \eta_{k1}\eta_{k3} = \sum_{k=1}^{K} \eta_{k2}\eta_{k3} = 0$. In general, the restricted $RC(M) + RL(M) + CL(M)$ model with consistent score restrictions in row, column, and layer scores across all dimensions has $IJK - I - J - K + 2 - M(I + J + K - 3) + M (M - 1)/2$ df. For some restricted $RC(M) + RL(M) + CL(M)$ models with consistent score restrictions on a subset but not all dimensions, it is even possible that *no* cross-dimensional constraint is required. For example, in the case of the $RC(2) + RL(2) + CL(2)$ model again, there is no need to impose any orthogonal restriction if consistent score restrictions in row, column, and layer scores are applied to the first dimension only.

Identifying Restrictions and Degrees of Freedom

The above discussion poses an interesting but seldom addressed issue on how to calculate the exact degrees of freedom of a given model. In general, the number of cross-dimensional constraints can be determined by calculating the rank of the Jacobian (Goodman, 1974; Siciliano & Mooijaart, 1997). For most practical purposes, it is probably easier to follow the empirical steps suggested by Wong (2001) to determine whether some if not all cross-dimensional constraints can be relaxed. As illustrated in Chapter 2, the degrees of freedom of a model equals the total number of cells minus the number of unique parameters, though the latter depends on the number of additional identifying restrictions as well. For example, when both centering and scaling restrictions are applied to the row scores for Mth-dimensional association models, only $(I - 2)M$ row score parameters are uniquely identified. On the other hand, if additional cross-dimensional restrictions are required for identification, the number of unique parameters becomes $(I - 2)M - M(M - 1)/2$. The same rule would apply to the column and layer score parameters as well, if any.

To determine whether certain identifying restrictions are required, it is important to recognize that the purpose of such restrictions is to solve the problem of numerical indeterminacy and to ensure that the converged estimates are unique. This is analogous to *underidentified* models in covariance structure analysis, that is, when the number of unknown parameters is greater than the number of available variances and covariances, infinite solutions can satisfy the same set of equations. The imposition of identifying restrictions and the lack thereof should have no impact on the goodness-of-fit statistics. The empirical steps outlined by Wong (2001), detailed below, use this property to verify whether all possible, some, or no cross-dimensional restrictions are needed and account for the correct degrees of freedom. The proposed strategy is needed for restricted multidimensional conditional association models because many of them require no or fewer restrictions than expected.

What follows is an outline of steps to determine whether cross-dimensional restrictions are necessary, but the procedures apply to other restrictions just as well:

(a) First, estimate the model without cross-dimensional constraints in the iterative stage, record the log-likelihood goodness-of-fit statistic $\left(L_1^2\right)$ and parameter estimates ($\tilde{\beta}_1$).

(b) Reestimate the model again with different (random) start values and record the log-likelihood goodness-of-fit statistic $\left(L_1^2\right)$ and parameter estimates ($\hat{\beta}_2$). If the goodness-of-fit statistics and parameter estimates remain the same, no cross-dimensional constraint is needed, and the estimates are unique. However, if only the fit statistics remain the same while the parameter estimates differ, it means that *some* cross-dimensional constraints are needed.

(c) Compile a list of potential cross-dimensional restrictions and add only one restriction in the iterative stage and compare the converged goodness-of-fit statistic, L_3^2, with L_1^2. If $L_1^2 = L_3^2$, this particular cross-dimensional restriction is needed. Otherwise, proceed with another restriction until all potential restrictions have been tested.

(d) Based on the results from Step (c), include two "required" restrictions simultaneously, and then compare the test statistics L_4^2 with L_1^2. If the results are identical, both restrictions are required. Proceed to incorporate additional restrictions incrementally until the list is completely exhausted. The incremental procedure is necessary because sometimes the cross-dimensional restrictions can be imposed individually on row, column, or layer score parameters but not simultaneously.

(e) Finally, adjust the degrees of freedom according to the number of valid cross-dimensional restrictions obtained from Step (d).

Example 3.1: Conditional Independence With Association Model

Table 3.1 presents a three-way cross-classification between individual political view (POLVIEWS), attitude toward gender and work (FEFAM), and attitude toward national welfare spending (NATFARE). Note that POLVIEWS, FEFAM, and NATFARE represent the row, column, and layer variables, respectively. The table is tabulated from the *2006 General Social Survey* and has 926 respondents. POLVIEWS is a self-identified political orientation measure and has seven categories: (1) extremely liberal, (2) liberal, (3) slightly liberal, (4) moderate, (5) slightly conservative, (6) conservative, and (7) extremely conservative. FEFAM measures the degree of agreement with the statement that "man is to work outside the home and woman is to take care of the home and family." The responses have been reorganized to strongly disagree, disagree, agree, and strongly agree. Finally, NATFARE has three responses regarding national welfare spending as too little, about right, and too much. This is an extension of an earlier example. In Chapter 2, we already discovered that the more politically conservative (or liberal) a person is, the more likely that she or he would agree (or disagree) with the "separate spheres" ideology. Our interest here is to further understand their interrelationship with attitude toward national welfare spending as well.

As expected, the model of complete independence (Line 1 of Table 3.2) does not fit the data well. The model has 74 *df*, L^2 of 186, and with slightly more than 16% misclassification. On the other hand, the relative goodness-of-fit of models with some or all two-way interaction parameters is significantly better than the complete independence model. With 36 *df*, the model with *FI* (Line 2) has L^2 of 35.35 and $p < .50$ and is clearly preferable. The next three models (Lines 3 to 5) are different specifications of conditional

Table 3.1 Cross-Classification of Political Views, Attitude Toward Gender and Work, and Welfare Spending

	National Welfare Spending (NATFARE)											
	Too Little				About Right				Too Much			
	Man to work outside the home and woman to take care of the home and family (FEFAM)											
Political views (POLVIEWS)	SD	D	A	SA	SD	D	A	SA	SD	D	A	SA
Extremely liberal	9	5	5	1	1	6	5	1	2	2	2	1
Liberal	17	13	7	4	13	22	9	1	7	13	6	2
Slightly liberal	8	14	6	0	10	29	10	0	5	14	6	2
Moderate	20	38	24	8	23	72	34	10	17	67	36	12
Slightly conservative	4	21	12	4	7	30	9	1	9	19	14	2
Conservative	2	9	8	3	1	16	19	2	11	28	28	11
Extremely conservative	0	1	5	0	2	3	3	2	2	7	6	6

SOURCE: *2006 General Social Survey.*

NOTE: SD, strongly disagree; D, disagree; A, agree; and SA, strongly agree.

Table 3.2 Analysis of Partial Association as Applied to Table 3.1

Model Description	df	L^2	BIC	Δ	p
1. Complete independence	72	167.59	−324.24	14.13	.000
2. Full two-way interaction	36	35.35	−210.56	5.72	.499
3. Conditional independence on POLVIEWS	42	47.25	−239.65	7.30	.267
4. Conditional independence on FEFAM	48	87.33	−240.56	10.33	.001
5. Conditional independence on NATFARE	54	91.04	−277.83	10.08	.001
6. $RC(1) + RL(1)$ partial association	57	68.58	−320.78	8.83	.140
7. Model 6 plus consistent row (POLVIEWS) score restrictions	62	72.77	−350.74	9.07	.165
8. Model 6 plus consistent and equality restrictions on row (POLVIEWS) scores ($\mu_1 = \mu_2, \mu_4 = \mu_5$)	64	73.59	−363.58	9.21	.193

NOTE: POLVIEWS, FEFAM, and NATFARE are treated as the row, column, and layer variables, respectively.

independence. The results suggest that, conditional on individual political orientation (POLVIEWS), there is independence between FEFAM and NAT-FARE (Line 3). The model has 42 df, L^2 of 47, and the observed relationship between FEFAM and NATFARE appears to be spurious when POLVIEWS is controlled. They both are determined by individual political orientation.

The contrast between Models 2 and 3 yields 6 df and L^2 of 11.90, the chi-square statistic is marginally significant ($p = .06$), indicating that there is relatively little to be gained to include interaction between FEFAM and NATFARE. Line 6 further decomposes the conditional independence parameters of the model in Line 3 using $RC(1)$ and $RL(1)$ partial association parameters (see Equation 3.8 for details). Although the relative goodness-of-fit of the present model deteriorates slightly, the deterioration of fit is statistically insignificant (15 df, $\Delta L^2 = 21.33$, $p = .12$) and is preferred instead. Finally, the model of Line 7 imposes consistent row score restrictions on the POLVIEWS variable (i.e., $\mu_i = \mu_i^*$), whereas that of Line 8 imposes additional inequality constraints such that $\mu_1 = \mu_2$ and $\mu_4 = \mu_5$, that is, the distance is the same between extremely liberal and liberal and between moderate and slightly conservative. The latter constraints help maintain the monotonic relationships found in the $RC(1)$ and $RL(1)$ partial associations. In both cases, the deterioration of fit is marginal but the last model (Line 8) provides the simplest understanding of the relationship involved among these three variables.

Table 3.3 reports the parameter estimates and their asymptotic standard errors of the last three models in Table 3.2. Since the point estimates from these models differ only slightly, the following discussion focuses mainly on Model 8 (i.e., with consistent row score and inequality restrictions). The *CIA* model informs us that not only is the relationship between FEFAM and NATFARE conditionally independent given POLVIEWS, but we can also adequately describe the relationship in highly parameterized forms as well. Generally speaking, the more liberal a person's political orientation is, the more likely that she or he would disagree with the "gendered" division of labor and would think that the American government has not been spending enough on the national welfare programs. The relationship between gendered division of labor and welfare spending, however, is spurious. Any observed relationship between the two is largely the outcome of individual political orientation instead. Furthermore, these partial relationships tend to be monotonic, with the exception that there is not much difference between those who are extremely liberal and those who are liberal and between those who are moderate and those who are slightly conservative. In retrospect, we could have imposed the restriction that $\mu_1 = \mu_2$ and/or $\mu_4 = \mu_5$ for the RC model in Table 2.4 earlier to achieve parsimonious understanding of the relationship between POLVIEWS and FEFAM.

Example 3.2: Partial Association Models

Table 3.4 presents a cross-classification of three indicators of satisfaction with life. They are satisfaction with family, family residence, and hobbies that correspond to row, column, and layer variables, respectively. Each variable has been recoded to have 4 categories: 1 for a fair amount, some, a little, or none; 2 for quite a bit, 3 for a great deal; and 4 for a very great deal. It has a total of 1,509 individuals. The table is derived from the *1977 General Social Survey* and has been analyzed extensively by Clogg (1982b). The reanalysis below will provide new and better understanding of the relationship involved among the three variables.

The complete independence (*I*) model (Line 1 of Table 3.5) yields 54 *df* and L^2 of 544 and clearly indicates that the three indicators of satisfaction with life are related to each other in a systematic fashion. The models of

Table 3.3 Selected Partial Association Parameters (POLVIEWS × FEFAM × NATFARE)

		Model 6	Model 7	Model 8
POLVIEWS × FEFAM **partial association**				
ϕ_{RC}		1.983 (0.744)	1.950 (0.374)	1.942 (0.374)
μ_i	Extremely liberal	−0.189 (0.175)	−0.413 (0.111)	−0.403 (0.043)
	Liberal	−0.456 (0.112)	−0.386 (0.083)	−0.403 (0.043)
	Slightly liberal	−0.365 (0.115)	−0.269 (0.084)	−0.279 (0.082)
	Moderate	−0.013 (0.082)	0.027 (0.057)	−0.002 (0.047)
	Slightly conservative	−0.063 (0.111)	−0.056 (0.077)	−0.002 (0.047)
	Conservative	0.421 (0.124)	0.494 (0.084)	0.492 (0.084)
	Extremely conservative	0.665 (0.110)	0.601 (0.086)	0.597 (0.087)
ν_j	Strongly disagree	−0.703 (0.064)	−0.726 (0.059)	−0.733 (0.058)
	Disagree	−0.208 (0.094)	−0.170 (0.093)	−0.162 (0.093)
	Agree	0.300 (0.110)	0.304 (0.111)	0.313 (0.111)
	Strongly agree	0.610 (0.092)	0.593 (0.095)	0.582 (0.097)

(Continued)

Table 3.3 (Continued)

POLVIEW × NATFARE partial association				
ϕ_{RL}		1.567	1.438	1.412
		(0.405)	(0.226)	(0.226)
μ_i	Extremely liberal	−0.606	−0.413	−0.403
		(0.124)	(0.111)	(0.043)
	Liberal	−0.275	−0.386	−0.403
		(0.115)	(0.083)	(0.043)
	Slightly liberal	−0.158	−0.269	−0.279
		(0.113)	(0.084)	(0.082)
	Moderate	0.065	0.027	−0.002
		(0.072)	(0.057)	(0.047)
	Slightly conservative	−0.051	−0.056	−0.002
		(0.100)	(0.077)	(0.047)
	Conservative	0.512	0.494	0.492
		(0.104)	(0.084)	(0.084)
	Extremely conservative	0.514	0.601	0.597
		(0.125)	(0.086)	(0.087)
η_k	Too little	−0.607	−0.580	−0.569
		(0.076)	(0.084)	(0.087)
	About right	−0.169	−0.207	−0.222
		(0.111)	(0.116)	(0.117)
	Too much	0.776	0.788	0.791
		(0.035)	(0.032)	(0.030)

NOTE: Values in parentheses are the asymptotic standard errors.

Table 3.4 Cross-Classification of Three Indicators of Satisfaction With Life

L =	R =	C = 1	C = 2	C = 3	C = 4
1	1	76	14	15	4
1	2	32	17	7	3
1	3	64	23	28	15
1	4	41	11	27	16
2	1	15	2	7	4
2	2	27	20	9	5
2	3	57	31	24	15
2	4	27	9	22	16
3	1	13	6	13	5
3	2	12	13	10	6
3	3	46	32	75	20
3	4	54	26	58	55
4	1	7	6	7	6
4	2	7	2	3	6
4	3	12	11	31	15
4	4	52	36	80	101

NOTE: Variables *L*, *R*, and *C* refer to satisfaction with hobbies, family, and family residence, respectively. Variable codes are 1 (*a fair amount, some, a little,* or *none*), 2 (*quite a bit*), 3 (*a great deal*), and 4 (*a very great deal*). This table has been analyzed by Clogg (1982b, Table 3) and is derived from the *1977 General Social Survey*.

Lines 2 through 5 reproduce a series of models reported by Clogg (1982b) to account for the relationship between the three indicators. Model 2 decomposes all two-way interaction parameters into one-dimensional partial association components (see Equation 3.10 for details). Even though the goodness-of-fit statistic of the current model has improved significantly over

Table 3.5 Analysis of Partial Association as Applied to Table 3.4

Model Description	df	L^2	BIC	Δ	p
1. Complete independence	54	544.37	149.13	23.73	.000
2. Unrestricted $RC(1) + RL(1) + CL(1)$	39	109.23	−176.22	9.81	.000
3. Restricted $RC(1) + RL(1) + CL(1)$ with consistent score restrictions	45	123.70	−205.66	10.90	.000
4. Unrestricted $RC(1) + RL(1) + CL(1)$ with consistent cells fitted exactly	35	37.68	−218.49	4.63	.347
5. Restricted $RC(1) + RL(1) + CL(1)$ with consistent score restrictions and consistent cells fitted exactly	41	49.15	−250.93	5.50	.179
6. Full two-way interaction	27	29.00	−168.62	4.86	.361
7. Full two-way interaction with consistent cells fitted exactly	23	21.93	−146.41	3.70	.524
8. Unrestricted uniform and log-multiplicative association in all partial association $U_{RC} + RC(1) + U_{RL} + RL(1) + U_{CL} + CL(1)$	36	45.85	−217.64	6.25	.126
9. Model (8) with consistent row scores	38	47.13	−231.00	6.25	.147
10. Model (8) with consistent column scores	38	50.90	−227.23	6.31	.079
11. Model (8) with consistent layer scores	38	53.63	−224.50	6.58	.048
12. Model (9) + $U_{RL} = U_{CL}$	39	48.57	−236.88	6.57	.140
13. Model (8) with consistent score restrictions	42	55.07	−252.34	6.69	.085
14. Model (13) + $U_{RL} = U_{CL}$	43	55.96	−258.77	6.82	.089
15. Model (14) − $CL(1)$	44	55.98	−266.06	6.83	.106

the complete independence model, it clearly still does not fit the data well (39 df, L^2 of 109). While the model with consistent score restrictions in row, column, and layer scores (Equation 3.11) appears to be quite consistent with the data, the overall fit of the latter model is nonetheless unsatisfactory.

Clogg (1982b) speculated that the major reason why the above models fail to provide satisfactory results is because of the four consistent cells in the cross-classifications, that is, (1,1,1), (2,2,2), (3,3,3), and (4,4,4) cells. In other words, individuals who consistently rated the three indicators of satisfaction with life as satisfactory or the lack thereof are quite different from the larger population. In fact, when all four cells have been blocked or fitted exactly, Clogg (1982b) was able to locate some association models that could best describe the relationship. They are now reproduced in Lines 4 and 5 here. The major difference between them is that the latter has consistent score restrictions in row, column, and layer scores across different partial associations in addition to the blocking of consistent cells. Although the goodness-of-fit statistics of both models (35 df, L^2 of 37.78 and 41 df, L^2 of 49.15, respectively) are satisfactory, we may want to step back and ask whether such treatment is indeed consistent with the data in the first place.

An implicit assumption in the blocking of the four consistent cells is the occurrence of three-way interaction between the three indicators of satisfaction of life involving those cells. To what extent does this assumption agree with existing data? The models of Lines 6 and 7 provide direct answers to this particular concern. Based on the test statistics, one cannot reject the *FI* model (Line 6) as the most preferred model than any of the previous ones discussed so far. It has 27 df and L^2 of 29 and is not statistically significant. Similarly, with the consistent cells fitted exactly, the reestimated *FI* model (Line 7) provides satisfactory results as well. Indeed, the contrast of the goodness-of-fit statistics between these two models indicates that, contrary to Clogg's (1982b) treatment, the four consistent cells are *not* the real culprit for the departure from the full two-way interaction. As a result, Clogg's preferred models (Lines 4 and 5) offer an incomplete picture at best and at worst misleading understanding of the relationship between the three indicators of life satisfaction.

To fully capture the complexity of all three partial association components, it is preferable to increase the dimensionality of the three partial association terms. Line 8 introduces two-dimensional association parameters by the inclusion of a uniform association component to each partial association. The model consumes 3 df resulting with 36 df and L^2 of 45.85. Statistically, the result is acceptable at the 0.10 level. The next three models (Lines 9 to 11) try to test whether some of the row, column, and layer scores are indeed consistent across different partial association components. The results suggest that only the consistent row scores restriction appears to be consistent with the data (Line 9). Conditional on model 9, Line 10 further imposes the restriction that the partial uniform association parameters between RL and CL are the same. The result appears to be affirmative.

Model 13 imposes the consistent score restrictions on row, column, and layer scores instead. Relative to Model 8, Model 13 has a gain of 6 df and ΔL^2 of 9.22 ($p = .16$) and is slightly more preferred. Based on a close inspection of the reported parameter estimates, it appears that the uniform association parameters in the RL and CL partial association can be equated. This hypothesis is then tested in Line 14, resulting in negligible change in the goodness-of-fit statistic. Finally, the estimates further suggest that there is no need to include the log-multiplicative component to account for the CL partial association. Instead, it can be adequately captured by a single uniform association parameter (U_{CL}) in the CL partial association compo-nent. The goodness-of-fit statistic of Line 15 again confirms that the hypothesis is indeed consistent with the data and cannot be rejected.

Thus, through the inclusion of two-dimensional hybrid (log-linear and log-multiplicative) partial association models, we end up with three alterna-tive specifications to understand the complex relationship involving the three indicators of satisfaction with life. They are quite different from the ones reported by Clogg (1982b) earlier. His preferred models are actually more complex than the ones presented here since there is little empirical support for three-way interaction among the three indicators. Of course, this does not mean that the patterns of partial association between the three indicators are simple either.

To fully comprehend the complex partial association patterns, the param-eter estimates of Models 9, 12, and 15 and their asymptotic standard errors are presented in Table 3.6. Since there are only minor differences between the estimates obtained from the three models, they portray rather similar relationships between the three indicators of satisfaction with life. All three models point to the importance of two-dimensional association patterns within each partial association, with the exception of Model 15 that the partial association between satisfaction with family residence and hobbies can be captured by a single partial uniform association parameter (U_{CL}). The partial intrinsic association between satisfaction with family and hob-bies ($\phi_{RL} = 1.1$) is particularly strong, relative to that between satisfaction with family and family residence ($\phi_{RC} = 0.6$). Although the parameter esti-mate of ϕ_{CL} is statistically significantly under Models 9 and 12, most of the estimated column row and score parameters, v_j and η_k, respectively, under the CL-partial association have large asymptotic standard errors and are not statistically significant. This underscores why the omission of the log-multiplicative RL-partial association in Model 15 does not lead to signifi-cant deterioration of fit.

Table 3.6 Selected Parameter Estimates of Partial Association Models as Applied to Table 3.4

		Model 9	Model 12	Model 15
RC partial association				
U_{RC}		0.134	0.133	0.133
		(0.025)	(0.025)	(0.025)
ϕ_{RC}		0.661	0.640	0.642
		(0.140)	(0.137)	(0.138)
μ_i	A fair amount	0.561	0.538	0.544
		(0.079)	(0.079)	(0.078)
	Quite a bit	−0.593	−0.598	−0.598
		(0.070)	(0.069)	(0.069)
	A great deal	−0.392	−0.389	−0.388
		(0.084)	(0.084)	(0.084)
	A very great deal	0.424	0.449	0.443
		(0.089)	(0.085)	(0.085)
ν_j	A fair amount	0.160	0.181	0.172
		(0.146)	(0.151)	(0.150)
	Quite a bit	−0.736	−0.742	−0.745
		(0.089)	(0.089)	(0.086)
	A great deal	−0.076	−0.081	−0.068
		(0.157)	(0.160)	(0.154)
	A very great deal	0.653	0.641	0.641
		(0.109)	(0.114)	(0.113)

(Continued)

Table 3.6 (Continued)

		Model 9	Model 12	Model 15
RL partial association				
U_{RL}		0.248	0.224	0.229
		(0.025)	(0.016)	(0.016)
ϕ_{RL}		1.137	1.140	1.138
		(0.140)	(0.137)	(0.137)
μ_i	A fair amount	0.561	0.538	0.544
		(0.079)	(0.079)	(0.078)
	Quite a bit	−0.593	−0.598	−0.598
		(0.070)	(0.069)	(0.069)
	A great deal	−0.392	−0.389	−0.388
		(0.084)	(0.084)	(0.084)
	A very great deal	0.424	0.449	0.443
		(0.089)	(0.085)	(0.085)
η_k	A fair amount	0.051	0.045	0.048
		(0.090)	(0.089)	(0.089)
	Quite a bit	−0.662	−0.662	−0.658
		(0.060)	(0.059)	(0.060)
	A great deal	−0.126	−0.122	−0.130
		(0.090)	(0.085)	(0.084)
	A very great deal	0.737	0.738	0.740
		(0.049)	(0.049)	(0.048)

		Model 9	Model 12	Model 15
CL partial association				
U_{CL}		0.203	0.224	0.229
		(0.023)	(0.016)	(0.016)
ϕ_{CL}		0.284	0.276	—
		(0.101)	(0.102)	
ν_j	A fair amount	−0.337	−0.278	—
		(0.301)	(0.308)	
	Quite a bit	−0.089	−0.140	—
		(0.361)	(0.370)	
	A great deal	0.841	0.848	—
		(0.089)	(0.080)	
	A very great deal	−0.415	−0.430	—
		(0.323)	(0.323)	
η_k	A fair amount	−0.090	0.017	—
		(0.361)	(0.368)	
	Quite a bit	−0.621	−0.652	—
		(0.240)	(0.230)	
	A great deal	0.776	0.749	—
		(0.152)	(0.180)	
	A very great deal	−0.065	−0.114	—
		(0.323)	(0.332)	

NOTE: The row, column, and layer variables are satisfaction with family, residence, and hobbies, respectively. Values in parentheses are the asymptotic standard errors.

Another important observation is that the estimated row, column, and layer scores (μ_i, v_j, and η_k, respectively) for the three indicators do not have a nice monotonic order. This explains why all those models that include only one-dimensional association component within each partial association fail to provide satisfactory fit. In the case of Clogg (1982b), he arbitrarily fixed the four consistent cells to achieve acceptable results. Consequentially, he erroneously concluded that there is a simple linear relationship among the three indicators. Unfortunately, the actual partial association patterns are far more complex. The simple adage that "the higher the level of satisfaction of life in one indicator is associated with higher satisfaction in another" is simply untenable.

CHAPTER 4. CONDITIONAL ASSOCIATION MODELS FOR THREE-WAY TABLES

For most social science applications, the analyses often include both two-way and three-way (and/or other higher-order) interaction terms. For instance, if the layer variable is a grouping variable, we are interested to know whether the association between row and column variables is the same across the entire range of the grouping variable. It should be reiterated that the term *grouping variable* is defined in the most generic form and may involve one or more variables. If the association between row and column variables is found to be different across the entire range of the grouping variable, we seek to further simplify our understanding by modeling substantively interpretable components, for example, using the linear or quadratic trend constraints and/or ANOVA-like decomposition method (Wong, 1995). On the other hand, if the layer variable is not a grouping variable and our interest lies in the complex three-way or higher-order interaction, then the three-mode association or related models may be more relevant (Anderson, 1996; Siciliano & Mooijaart, 1997; Tucker, 1966; see especially Wong, 2001, about the relationship between some restricted three-mode association models with several conditional association models discussed in this chapter). However, the latter will not be discussed due to space limitation.

Unlike the partial association models introduced in the previous chapter, relationships between row and layer and between column and layer variables are treated as uninteresting, and the corresponding parameters are therefore not decomposed. Instead, we are mainly interested in the association between row and column variables, conditional on the layer variable. Technically, the types of models introduced in this chapter are known as conditional association models. They offer interesting ways to capture departures from the conditional independence model and to locate specifically where group differences lie (note particularly the works by Becker, 1989a, 1989b, 1990; Becker & Clogg, 1989; Clogg, 1982a, 1982b; Clogg & Shihadeh, 1994; Erikson & Goldthorpe, 1992; Goodman & Hout, 1998, 2001; Wong, 1990, 1992, 1995; Xie, 1992; Yamaguchi, 1987).

Conditional Independence (*CI* or Conditional *RC*(0)) Model

Suppose there are three variables A, B, and C representing row, column, and layer, respectively, with variable C as the grouping variable. Similar to the exposition earlier in Chapter 3, the conditional independence (*CI*) model (compare Equation 3.3) can be written as follows:

$$\log F_{ijk} = \lambda + \lambda_i^A + \lambda_j^B + \lambda_k^C + \lambda_{ik}^{AC} + \lambda_{jk}^{BC}, \tag{4.1}$$

with $\lambda_{ij}^{AB} = \lambda_{ijk}^{ABC} = 0$. Under the current specification, the model postulates that after controlling for variable *C*, there is no relationship between variables *A* and *B*. In other words, the observed *AB* interaction in the aggregated table is largely *spurious* and caused by the common variable, *C*, instead. The conditional independence model has $(I-1)(J-1)K$ *df*. Since Equation 4.1 does not contain any log-multiplicative row and column effects components, it can be termed equivalently as the conditional *RC*(0) model as well. According to Chapter 3, the conditional local log-odds ratios for *A* and *B* given *C* and the ratio of the conditional local log-odds ratios under conditional independence have the same outcome:

$$\log \theta_{ij(k)} = \log \theta_{ijk} = 0. \tag{4.2}$$

Homogeneous or Constant Association Model

If we only impose the restriction that $\lambda_{ijk}^{ABC} = 0$, Equation 4.1 becomes the conventional log-linear model with all or full two-way interaction (or the *FI* model in Equation 3.9). The model is otherwise known as the homogeneous or constant association model because it assumes that the *unspecified* association pattern found between row and column variables does not differ across layers. Stated differently, the model postulates constant odds ratios across layers, and it can be written as follows:

$$\log F_{ijk} = \lambda + \lambda_i^A + \lambda_j^B + \lambda_k^C + \lambda_{ik}^{AC} + \lambda_{jk}^{BC} + \lambda_{ij}^{AB}, \tag{4.3}$$

and the corresponding conditional local odds ratios and the ratios of the conditional local odds ratios are as follows:

$$\log \theta_{ij(k)} = \log \theta_{ij}, \tag{4.4}$$

and

$$\log \theta_{ijk} = 0,$$

where $i = 1, \ldots, I$, $j = 1, \ldots, J$, and $k = 1, \ldots, K$. The model has $(I-1)(J-1)(K-1)$ *df*.

Three-Way Interaction or Saturated Model

Finally, when both λ_{ij}^{AB} and λ_{ijk}^{ABC} terms are included, the model becomes fully saturated with 0 df. The three-way interaction or saturated model can be represented as follows:

$$\log F_{ijk} = \lambda + \lambda_i^A + \lambda_j^B + \lambda_k^C + \lambda_{ik}^{AC} + \lambda_{jk}^{BC} + \lambda_{ij}^{AB} + \lambda_{ijk}^{ABC}. \qquad (4.5)$$

Researchers in the past often faced an empirical dilemma when they were forced to choose between Equations 4.3 and 4.5. On the one hand, according to conventional theory of statistical testing, the goodness-of-fit statistic relative to its degrees of freedom for the constant association model usually does not fit well, and the saturated model would therefore be preferred. This is especially true when the sample size is relatively large. On the other hand, other model selection strategies such as *BIC* and *AIC* tend to prefer the constant association model using the principle of scientific parsimony. In reality, researchers recognize that both log-linear models are *wrong* and choosing one wrong model over another (probably slightly less) wrong one does not seem to be the correct or defensible strategy. Since the "true" model lies somewhere between the two, we need to develop intermediate statistical models that are powerful enough to locate cross-group differences and, better yet, to separate and distinguish *patterns* (structures) from *levels* (group differences) of association. For example, it would be interesting to uncover that while the set of odds ratios may have a particular *structure* that is common across the grouping variable, the *levels* of association can differ across groups.

Layer Effect Models to Model Group Differences

The first set of models to detect group differences are known as layer effect models (Erikson & Goldthorpe, 1992; Goodman & Hout, 1998, 2001; Wong, 1990, 1992; Xie, 1992; Yamaguchi, 1987). They share a common feature in the use of statistically powerful 1-df test per layer to detect group differences. In most cases, they often do not postulate an explicit structure or pattern of association. The latter observation, however, is more of a practical application limitation rather than an inherent deficiency (see Goodman & Hout, 1998; Xie, 1998; Yamaguchi, 1998, for more in-depth discussion). The only major difference between them lies in their specifications on how the layers differ from each other.

The log-linear layer effect model (LL_1; Wong, 1990; Yamaguchi, 1987) is the first formulated intermediate statistical model proposed to detect group differences. It assumes that both row and column variables possess ordinal properties or, statistically speaking, the association patterns between row and column variables are *isotropic*.[1] The model is also known as the log-linear uniform difference model (Wong, 1994) and can be written as follows:

$$\log F_{ijk} = \lambda + \lambda_i^A + \lambda_j^B + \lambda_k^C + \lambda_{ik}^{AC} + \lambda_{jk}^{BC} + \lambda_{ij}^{AB} + \beta_k U_i V_j \qquad (4.6)$$

The model has $(I - 1)(J - 1)(K - 1) - (K - 1) = (IJ - I - J)(K - 1)$ *df.* Under Equation 4.6, not all β_k parameters can be uniquely identified. The conventional normalization is to set $\beta_1 = 0$. Under such normalization, β_k represents the deviation in the common set of odds ratios for layer k from layer 1 (the baseline group). The difference between the log-likelihood test statistics of Equations 4.3 and 4.6 yields a chi-square statistic with $K - 1$ *df.* This statistically powerful test allows researchers to affirm the existence of group differences. Based on Equation 4.6, the log-linear layer effect model postulates the following relationship among the conditional local odds ratios:

$$\log \theta_{ij(k)} = \log \theta_{ij} + \beta_k. \qquad (4.7)$$

Thus, the complete set of odds ratios for each group differs by a multiplicative scale factor, β_k. From Equation 4.7, it can be shown easily that when comparing layer k versus layer k',

$$\log \theta_{ij(k)} - \log \theta_{ij(k')} = \beta_k - \beta_{k'}, \qquad (4.8)$$

and

$$\frac{\log \theta_{ij(k)}}{\log \theta_{ij(k')}} = \frac{\log \theta_{ij} + \beta_k}{\log \theta_{ij} + \beta_{k'}}. \qquad (4.9)$$

In other words, the differences between conditional log-odds ratios can be simplified into uniform differences, whereas the ratios of the conditional log-odds ratios cannot be reduced into simpler terms. Note that the former is equivalent to the local odds ratios defined in Equation 3.3. The relationship described in Equation 4.8 also explains why it is known as the log-linear uniform difference model. Furthermore, the model is closely related to the uniform association model discussed in Chapter 2.

Instead of multiplicative scale or log-additive differences between layers, the second-layer effect model postulates that the differences are log-multiplicative instead and becomes the log-multiplicative layer effect model (LL_2), also widely known as the UNIDIFF model in comparative mobility literature (Erikson & Goldthorpe, 1992; Xie, 1992; Xie & Pimentel, 1992). Unlike the previous formulation, it does not impose any ordinal restriction on the row and column categories and can be written as follows:

$$\log F_{ijk} = \lambda + \lambda_i^A + \lambda_j^B + \lambda_k^C + \lambda_{ik}^{AC} + \lambda_{jk}^{BC} + \phi_k \psi_{ij}, \qquad (4.10)$$

where ψ_{ij} represents the full interaction between variables A and B. Note that ψ_{ij} parameters are identified by the restrictions that $\sum_i \psi_{ij} = \sum_j \psi_{ij} = 0$, and ϕ_k parameters are identified by the restriction $\sum_k \phi_k^2 = 1$. The model has $(I-1)(J-1)(K-1) - (K-1) = (IJ - I - J)(K-1)$ df. Similar to the case of the log-linear layer effect model, the difference between the log-likelihood test statistics of Equations 4.3 and 4.10 also yields a chi-square statistic with $K - 1$ df to test for group differences. Instead of an unstructured interaction pattern, it is possible to impose explicit structure or pattern for ψ_{ij}, say topological model, row effects model, or alike, to represent AB association (Xie, 1992). Under the above specification, the conditional log-odds ratios can be written as follows:

$$\log \theta_{ij(k)} = \phi_k \log \theta_{ij}. \qquad (4.11)$$

In other words, the complete set of odds ratios for each group differs by a log-multiplicative scale factor, ϕ_k. From Equation 4.11, it can be shown easily that when comparing layer k versus layer k', the differences of the conditional log-odds ratios and the ratios of the conditional log-odds ratios have the following relationships:

$$\log \theta_{ij(k)} - \log \theta_{ij(k')} = (\phi_k - \phi_{k'})\log \theta_{ij}, \qquad (4.12)$$

and

$$\frac{\log \theta_{ij(k)}}{\log \theta_{ij(k')}} = \frac{\phi_k}{\phi_{k'}}. \qquad (4.13)$$

Equations 4.12 and 4.13 indicate that the differences of the conditional log-odds ratios cannot be reduced into simpler terms whereas the ratios of the conditional log-odds ratio for pairs of cells in one layer, k, and the corresponding conditional log-odds ratio for the same pairs in another layer, k', is constant. This confirms that the difference between layers is log-multiplicative rather than log-additive in nature.

If neither the log-linear nor the log-multiplicative layer effect models provide satisfactory results, the third-layer effect specification postulates an even more complex formulation known as the modified regression-type layer effect model (LL_3; Goodman & Hout, 1998, 2001). The model bears strong resemblance to the log-multiplicative layer effect model above, except that it includes λ_{ij}^{AB} term as well and can be written as follows:

$$\log F_{ijk} = \lambda + \lambda_i^A + \lambda_j^B + \lambda_k^C + \lambda_{ik}^{AC} + \lambda_{jk}^{BC} + \lambda_{ij}^{AB} + \phi_k \psi_{ij}. \qquad (4.14)$$

The model has $(IJ - I - J)(K - 2)$ df. According to Goodman and Hout (1998), the conditional log-odds ratios can be written as follows:

$$\log \theta_{ij(k)} = \zeta_{ij} + \zeta_{ij}' \phi_k, \qquad (4.15)$$

where

$$\zeta_{ij} = \lambda_{ij}^{AB} + \lambda_{i+1,j+1}^{AB} - \lambda_{i+1,j}^{AB} - \lambda_{i,j+1}^{AB}, \qquad (4.16)$$

and

$$\zeta_{ij}' = \psi_{ij} + \psi_{i+1,j+1} - \psi_{i+1,j} - \psi_{i,j+1}. \qquad (4.17)$$

The above specification bears strong resemblance with ordinary least squares (OLS) specification where $E(y|x) = \beta_0 + \beta_1 x$. Under the above formulation, when comparing layer k versus layer k', the differences of the conditional log-odds ratios and the ratios of the conditional log-odds ratios can be written as follows:

$$\log \theta_{ij(k)} - \log \theta_{ij(k')} = \zeta_{ij}'(\phi_k - \phi_{k'}), \qquad (4.18)$$

and

$$\frac{\log \theta_{ij(k)}}{\log \theta_{ij(k')}} = \frac{\zeta_{ij} + \zeta_{ij}' \phi_k}{\zeta_{ij} + \zeta_{ij}' \phi_{k'}}. \qquad (4.19)$$

From Equations 4.18 and 4.19, neither the differences of the conditional log-odds ratios nor the ratios of the conditional log-odds ratios are uniform across different i, j. Rather, the ratios of cross-layer differences in the log-odds ratios are all proportional. That is,

$$\frac{\log \theta_{ij(k)} - \log \theta_{ij(k')}}{\log \theta_{ij(k)} - \log \theta_{ij(k^*)}} = \frac{\phi_k - \phi_{k'}}{\phi_k - \phi_{k^*}}, \qquad (4.20)$$

for $k \neq k' \neq k*$.[2] The relative merits of the regression-type approach include (1) whereas λ_{ij}^{AB} establishes the baseline pattern of association, ψ_{ij} and ϕ_k adjust that baseline pattern for layer k, and (2) the baseline pattern of association, which is established by λ_{ij}^{AB}, may or may not apply to a particular layer, depending on the parameterization of ψ_{ij} and ϕ_k. For instance, one can use full interaction, uniform association, RC association, topological models, or alike constraints either to λ_{ij}^{AB}, ψ_{ij}, or both (note particularly the comments by Xie, 1998, and Yamaguchi, 1998, about the merits of the latter specifications). To generate meaningful cross-layer comparisons, the minimum number of layers under LL_3 would be 3.

Association Models to Model Group Differences

The above layer effects models offer statistically powerful tests to detect differences across layers when they provide satisfactory goodness-of-fit statistics relative to their degrees of freedom. If they do not, it is possible to extend the family of association models developed from earlier chapters to model group differences as well. Similar to our discussion earlier, the latter can be broadly grouped into two specific types: (1) whether the structure of association can be represented in log-linear terms, log-multiplicative terms, or both (i.e., hybrid models) and (2) the degree of complexity in association as one- or multidimensional (Becker, 1989a; Becker & Clogg, 1989; Wong, 2001). Of course, hybrid association models with both log-linear and log-multiplicative components are by construction at least two dimensional.

Log-Linear Specifications: (R+C)-L Models

Rather than a full exposition of the application of various log-linear formulations in multiple tables, the discussion below focuses on the log-linear row and column effects $(R+C)$ specification instead. The discussion can be generalized easily to other models such as U, R, and C, but they will not be discussed because of space limitation.[3] As it will be illustrated shortly, the application of the log-linear row and column effects model $(R+C)$ to multiple tables closely parallels its log-multiplicative (RC) counterpart.

Consistent with the terminology adopted by Becker (1989a) and Becker and Clogg (1989), the class of log-linear row and column effects model in multiple tables would be termed as $(R+C)$-L models, where the first half of the term corresponds to the type of association models involved, and the latter half corresponds to conditional association with the layer variable. As usual, the letters R, C, and L represent the row, column, and layer variables, respectively. The simplest case under such specification would be the homogeneous log-linear rows and column effects model (homogeneous (*R+C*)-*L* or simply homogeneous *R+C*). Note that the term *homogeneous* refers to equal row and column effects across all levels of the layer variable, *C*.[4] Likewise, the term *heterogeneous* refers to layer-specific row and column effects. Formally, the homogeneous model can be represented as follows:

$$\log F_{ijk} = \lambda + \lambda_i^A + \lambda_j^B + \lambda_k^C + \lambda_{ik}^{AC} + \lambda_{jk}^{BC} + \phi U_i V_j + \tau_i^A V_j + \tau_j^B U_i, \quad (4.21)$$

where U_i and V_j are fixed integer row and column scores, respectively. To identify both τ_i^A and τ_j^B parameters, the following normalizations are needed: $\tau_1^A = \tau_I^A = \tau_1^B = \tau_J^B = 0$. The model therefore has $(I-1)(J-1)K - (I-2) - (J-2) - 1 = IJK - (K+1)(I+J-1) + 2 \; df$. This is a special case of the constant association model earlier (Equation 4.3), because it postulates no 3-factor interaction or no layer differences in association. The partial or conditional odds ratios $\theta_{ij(k)}$ and the local odds ratios θ_{ijk} can be written as follows:

$$\log \theta_{ij(k)} = \phi + \left(\tau_{i+1}^A - \tau_i^A\right) + \left(\tau_{j+1}^B - \tau_j^B\right), \quad (4.22)$$

and

$$\log \theta_{ijk} = \log \theta_{ij(k+1)} - \log \theta_{ij(k)} = 0. \quad (4.23)$$

Note that the above conditional odds ratios do not include $\log \theta_{i(j)k}$ or $\log \theta_{(i)jk}$, since they cannot be further simplified and include additional partial λ-parameters for *AC* and *BC* interaction.

If we permit both the association (ϕ) and row and column parameters (τ_i and τ_j) to vary across different levels of *k*, Equation 4.21 becomes the heterogeneous model (with 3-factor interaction). The heterogeneous (*R+C*)-*L* or *R+C* model can be represented as follows:

$$\log F_{ijk} = \lambda + \lambda_i^A + \lambda_j^B + \lambda_k^C + \lambda_{ik}^{AC} + \lambda_{jk}^{BC} + \phi_k U_i V_j + \tau_{ik}^{AC} V_j + \tau_{jk}^{BC} U_i. \quad (4.24)$$

The following normalizations can be adopted: $\tau_{11}^A = \cdots = \tau_{1K}^A = \tau_{I1}^A = \cdots = \tau_{IK}^A = \tau_{11}^B = \cdots = \tau_{1K}^B = \tau_{J1}^B = \cdots = \tau_{JK}^B = 0$ and has $(I-1)(J-1)K - (I-2)K - (J-2)K - K = (I-2)(J-2)K \; df$. Under the current formulation, the partial or conditional odds ratios $\theta_{ij(k)}$ and the local odds ratios θ_{ijk} can be written as follows:

$$\log \theta_{ij(k)} = \phi_k + \left(\tau_{i+1,k}^{AC} - \tau_{ik}^{AC}\right) + \left(\tau_{j+1,k}^{BC} - \tau_{jk}^{BC}\right), \quad (4.25)$$

and

$$\log\theta_{ijk} = (\phi_{k+1} - \phi_k) + \left(\tau^{AC}_{i+1,k+1} + \tau^{AC}_{ik} - \tau^{AC}_{i+1,k} - \tau^{AC}_{i,k+1}\right)$$
$$+ \left(\tau^{BC}_{j+1,k+1} + \tau^{BC}_{jk} - \tau^{BC}_{j+1,k} - \tau^{BC}_{j,k+1}\right). \tag{4.26}$$

Again, both $\log\theta_{i(j)k}$ and $\log\theta_{(i)jk}$ involve partial λ-parameters for AC and BC interaction as well as log-linear row and column effects components and cannot be further simplified.

When one of the above models offers satisfactory result, a comparison of the goodness-of-fit statistics between them provides valuable information about the total variation in rows and column effects. Following the logic of the layer effect models developed previously, it is possible to formulate intermediate models that capture layer differences with 1-*df* test. Using the terminology adopted by Becker (1989a) and Becker and Clogg (1989), these models are termed *partial heterogeneous* or *partial homogeneous* (*R+C*)-*L* model. In particular, a model that postulates homogeneous row and column effects parameters across layers in the last two components of Equation 4.24 but layer-specific effects in ϕ_k is of special interests and can be specified as follows:

$$\log F_{ijk} = \lambda + \lambda^A_i + \lambda^B_j + \lambda^C_k + \lambda^{AC}_{ik} + \lambda^{BC}_{jk} + \phi_k U_i V_j + \tau^A_i V_j + \tau^B_j U_i, \tag{4.27}$$

with $IJK - (I + J)(K + 1) + 4$ *df*. Under this formulation, the conditional and local odds ratios can be written, respectively, as the following:

$$\log\theta_{ij(k)} = \phi_k + \left(\tau^A_{i+1} - \tau^A_i\right) + \left(\tau^B_{j+1} - \tau^B_j\right), \tag{4.28}$$

and

$$\log\theta_{ijk} = \log\theta_{ij(k+1)} - \log\theta_{ij(k)} = \phi_{k+1} - \phi_k. \tag{4.29}$$

In other words, the above model transfers all layer differences to the single ϕ_k parameters while maintaining row and column effects parameters constant. Constructed in this manner, it is clear that the above model is a special case of the log-linear layer effect model discussed earlier. Whereas the last two terms of Equation 4.28, that is, differences in the log-linear row and column effects, represents common *pattern* or *structure* of association, the term ϕ_k represents different *levels* of association. Other partial homogeneous models are feasible as well. For example, the model that postulates that only the column effects are constant across the grouping variable has the following form:

$$\log F_{ijk} = \lambda + \lambda_i^A + \lambda_j^B + \lambda_k^C + \lambda_{ik}^{AC} + \lambda_{jk}^{BC} + \phi_k U_i V_j + \tau_{ik}^{AC} V_j + \tau_j^B U_i, \quad (4.30)$$

whereas the model that postulates that only the row effects are constant across the grouping variable has a similar formulation:

$$\log F_{ijk} = \lambda + \lambda_i^A + \lambda_j^B + \lambda_k^C + \lambda_{ik}^{AC} + \lambda_{jk}^{BC} + \phi_k U_i V_j + \tau_i^A V_j + \tau_{jk}^{BC} U_i. \quad (4.31)$$

Note that in both cases, ϕ_k also covary because it would be illogical by just permitting the row or column effects parameters to vary without allowing the level of association (ϕ_k) to do so as well. Relative to Equation 4.27, Equation 4.30 uses an additional $(I - 2)(K - 1)$ df, whereas Equation 4.31 uses an additional $(J - 2)(K - 1)$ df. As a result, the model specified under Equation 4.30 has $(J - 2)(IK - K - 1)$ df, while the one under Equation 4.31 has $(I - 2)(JK - K - 1)$ df instead.

When the model of Equation 4.27 fits the data well and the grouping variable indexes temporal order such as survey years or birth cohorts, it is possible to postulate that the intrinsic association parameters (ϕ_k) exhibit decreasing, increasing, or nonlinear trends as well (Wong, 1995; Wong & Hauser, 1992). For instance, models with linear and quadratic trend restrictions on ϕ_k will take the following forms, respectively:

$$\phi_k = \phi(1 + a\,t), \quad (4.32)$$

and

$$\phi_k = \phi(1 + a\,t + b\,t^2). \quad (4.33)$$

By comparing the above two specifications with their homogeneous and heterogeneous counterparts, one can use differences in chi-square statistics to determine whether ϕ_k parameters have indeed changed over time or across birth cohorts. Of course, if there are enough observational units in the layer variable, it is even possible to include other parametric specifications such as the spline-like regression function as well.

If the layer variable involves two or more variables such as gender and ethnicity, it is possible to decompose ϕ_k parameters into ANOVA-like components (Raymo & Xie, 2000; Wong, 1995). For the sake of convenience, suppose that there are only two layer variables L_1 and L_2 with M and N

categories, respectively. The fully interactive model for ϕ_k under Equation 4.27 is equivalent to the following:

$$\phi_k = \phi_{mn}, \tag{4.34}$$

whereas the homogeneous $(R+C)$-L model postulates that

$$\phi_k = \phi. \tag{4.35}$$

Several intermediate formulations are of special interest:

$$\phi_k = \phi_m + \phi_n. \tag{4.36}$$

$$\phi_k = \phi_m. \tag{4.37}$$

$$\phi_k = \phi_n. \tag{4.38}$$

Instead of a fully interactive representation of ϕ_k in Equation 4.34, the specifications under Equations 4.36 to 4.38 represent additive restrictions. They differ from each other in terms of the number of those additive component(s) involved. A comparison of their goodness-of-fit statistics informs the possibility whether the intrinsic association parameters can be represented in simpler terms. Of course, when the layer variable indexes both temporal order and multiple groups (say, ethnicity by gender or country by gender), it is possible to impose both types of constraints simultaneously to locate the sources of cross-table variation. Note that these constraints can be applied to all previous layer effect models or any 1-df tests involving ϕ_k or ϕ_{mk} in one- or multidimensional RC association models to be discussed shortly.

Log-Multiplicative Specifications: RC(M)-L Models

By the same token, one can extend the above strategy to incorporate log-multiplicative row and column components as conditional association models. For exposition purposes, the discussion below will begin with the most general case, multidimensional RC conditional association models, $RC(M)$-L. Using the same nomenclature adopted earlier, the first half of the term, $RC(M)$, corresponds to multidimensional RC components, whereas the second half denotes that the model is conditional on the layer variable, L. The appropriate modeling strategy is to determine first the number of dimensions needed to understand departures from conditional independence, and then to locate the exact source of variation among different log-multiplicative components, if any, once the dimensionality has been determined.

We begin with two extreme models, the homogeneous $RC(M)$-L model and the heterogeneous $RC(M)$-L model. Note that the values for M can range from 0 to $\min(I - 1, J - 1)$, that is, $0 \leq M \leq \min(I - 1, J - 1)$. When $M = 0$, the model would be equivalent to the conditional independence model, and when M reaches its maximum in the homogeneous formulation, it would be equivalent to the full two-way or constant association model instead. The homogeneous $RC(M)$-L or simply homogeneous $RC(M)$ model can be represented as the following:

$$\log F_{ijk} = \lambda + \lambda_i^A + \lambda_j^B + \lambda_k^C + \lambda_{ik}^{AC} + \lambda_{jk}^{BC} + \sum_{m=1}^{M} \phi_m \mu_{im} \nu_{jm}. \quad (4.39)$$

The model has $(I - 1)(J - 1)K - M(I + J - M - 5)$ df. To uniquely identify all μ_{im} and ν_{jm} parameters, centering, scaling, and cross-dimensional constraints are required. In other words, the following constraints are needed: $\sum_{i=1}^{I} \mu_{im} = \sum_{j=1}^{J} \nu_{jm} = 0$ and $\sum_{i=1}^{I} \mu_{im} \mu_{im'} = \sum_{j=1}^{J} \nu_{jm} \nu_{jm'} = \delta_{mm'}$, where $\delta_{mm'}$ is the Kronecker delta (i.e., $\delta_{mm'} = 1$ when $m = m'$ and 0 otherwise).

Under Equation 4.39, the conditional log-odds ratios and the ratios of the conditional log-odds ratios in the homogeneous $RC(M)$-L model can be written as follows:

$$\log \theta_{ij(k)} = \sum_{m=1}^{M} \phi_m (\mu_{i+1,m} - \mu_{im})(\nu_{j+1,m} - \nu_{jm}), \quad (4.40)$$

and

$$\log \theta_{ijk} = 0. \quad (4.41)$$

On the other hand, one can formulate the heterogeneous $RC(M)$-L model such that all ϕ_m, μ_{im}, and ν_{jm} parameters would covary with the layer variable (L). The model can be written as follows:

$$\log F_{ijk} = \lambda + \lambda_i^A + \lambda_j^B + \lambda_k^C + \lambda_{ik}^{AC} + \lambda_{jk}^{BC} + \sum_{m=1}^{M} \phi_{mk} \mu_{imk} \nu_{jmk}. \quad (4.42)$$

The model has $(I - M - 1)(J - M - 1)K$ df. Similar to the previous case, centering, scaling, and cross-dimensional constraints on the row and column score parameters are needed to achieve unique identification.

Under Equation 4.42, the conditional log-odds ratios and the ratios of the conditional log-odds ratios of the heterogeneous $RC(M)$-L model can be written as follows:

$$\log \theta_{ij(k)} = \sum_{m=1}^{M} \phi_{mk} (\mu_{i+1,mk} - \mu_{imk})(\nu_{j+1,mk} - \nu_{jmk}), \quad (4.43)$$

and

$$\log\theta_{ijk} = \sum_{m=1}^{M} \phi_{m,\,k+1}(\mu_{i+1,\,m,\,k+1} - \mu_{im,\,k+1})(\nu_{j+1,\,m,\,k+1} - \nu_{jm,\,k+1})$$
$$- \sum_{m=1}^{M} \phi_{mk}(\mu_{i+1,\,mk} - \mu_{imk})(\nu_{j+1,\,mk} - \nu_{jmk}). \qquad (4.44)$$

Table 4.1 provides a summary of various types of $RC(M)$-L conditional association models. It has three columns: model specification, types of cross-dimensional constraints involved, and its associated degrees of freedom. Note that the numerical illustrations include only up to three-dimensional models, $RC(3)$-L, because there is generally little need to entertain further higher-order complex interactive patterns. In fact, most empirical data can be adequately analyzed by the two-dimensional RC conditional association models. In the event that such specification does not provide satisfactory results, it is possible to add simple but substantively interpretable parameters that yield satisfactory results (e.g., the prevalence of immobility in social mobility research). With the exception of $RC(1)$ models, *all* unrestricted $RC(M)$-L conditional association models require cross dimensional constraints.

By estimating a series of $RC(M)$ conditional association models, it is possible to use their goodness-of-fit statistics to decipher how many dimensions are needed to understand cross-layer differences in association (Becker & Clogg, 1989). This particular chi-square partitioning strategy will be discussed in the two illustrative examples shortly. After the number of RC dimensions needed has been determined, the next step is to formulate partial homogeneous or partial heterogeneous restrictions to systematically explore which components (ϕ_{mk}, μ_{imk}, and ν_{jmk}) vary across layers and which do not.

Since most empirical analyses involve no more than two dimensions in the log-multiplicative components, it would be more fruitful to concentrate the discussion in those occasions. When $M = 1$, the following partial homogeneous or partial heterogeneous $RC(1)$-L models are of interest:

(a) Homogeneous μ_i, heterogeneous ϕ and ν_j

$$\log F_{ijk} = \lambda + \lambda_i^A + \lambda_j^B + \lambda_k^C + \lambda_{ik}^{AC} + \lambda_{jk}^{BC} + \phi_k \mu_i \nu_{jk}. \qquad (4.45)$$

The model has $K(I-2)(J-2) + (K-1)(I-2) = (I-2)(JK-K-1)$ df. Note again that it does not make much substantive sense to just allow ν_j to

Table 4.1 Cross-Dimensional Constraints and Degrees of Freedom for Some $RC(M)$-L Models for $I \times J \times K$ Cross-Classification Table

Model	Cross-Dimensional Constraints	df
1. Homogeneous $RC(0)$	Not applicable	$(I-1)(J-1)K$
2. Heterogeneous $RC(1)$	Not applicable	$(I-2)(J-2)K$
3. Homogeneous $RC(1)$	Not applicable	$(I-1)(J-1)K - ((I+J-3)$
4. Heterogeneous $RC(2)$	$\sum \mu_{i1k}\mu_{i2k} = \sum \nu_{j1k}\nu_{j2k} = 0$ $\forall \, k = 1, 2, \ldots, K$	$(I-3)(J-3)K$
5. Homogeneous $RC(2)$	$\sum \mu_{i1}\mu_{i2} = \sum \nu_{j1}\nu_{j2} = 0$	$(I-1)(J-1)K - 2\,(I+J-4)$
6. Heterogeneous $RC(3)$	$\sum \mu_{i1k}\mu_{i2k} = \sum \nu_{j1k}\nu_{j2k} =$ $\sum \mu_{i1k}\mu_{i3k} = \sum \nu_{j1k}\nu_{j3k} =$ $\sum \mu_{i2k}\mu_{i3k} = \sum \nu_{j2k}\nu_{j3k} = 0$	$(I-4)(J-4)K$

88

Model	Cross-Dimensional Constraints	df
7. Homogeneous $RC(3)$	$\sum \mu_{i1} \mu_{i2} = \sum \nu_{j1} \nu_{j2} =$ $\sum \mu_{i1} \mu_{i3} = \sum \nu_{j1} \nu_{j3} =$ $\sum \mu_{i2} \mu_{i3} = \sum \nu_{j2} \nu_{j3} = 0$	$(I-1)(J-1)K - 3(I+J-5)$
8. Heterogeneous $RC(M)$	$\sum \mu_{imk} \mu_{im'k} = \sum \nu_{jmk} \nu_{jm'k} = 0$ where $m \neq m'$	$(I-M-1)(J-M-1)K$
9. Homogeneous $RC(M)$	$\sum \mu_{im} \mu_{im'} = \sum \nu_{jm} \nu_{jm'} = 0$ where $m \neq m'$	$(I-1)(J-1)K - M(I+J-M-5)$

vary across layers, while ϕ would be invariant. The conditional log-odds ratios and the local log-odds ratios, respectively, are as follows:

$$\log\theta_{ij(k)} = \phi_k(\mu_{i+1} - \mu_i)(v_{j+1,\,k} - v_{jk}), \qquad (4.46)$$

and

$$\log\theta_{ijk} = [\phi_{k+1}(v_{j+1,\,k+1} - v_{j,\,k+1}) - \phi_k(v_{j+1,\,k} - v_{jk})](\mu_{i+1} - \mu_i). \quad (4.47)$$

(b) Homogeneous v_j, heterogeneous ϕ and μ_i

$$\log F_{ijk} = \lambda + \lambda_i^A + \lambda_j^B + \lambda_k^C + \lambda_{ik}^{AC} + \lambda_{jk}^{BC} + \phi_k\mu_{ik}v_j. \qquad (4.48)$$

The model has $K(I - 2)(J - 2) + (K - 1)(J - 2) = (J - 2)(IK - K - 1)$ df and the conditional or partial log-odds ratios and the local log-odds ratios, respectively, are as follows:

$$\log\theta_{ij(k)} = \phi_k(\mu_{i+1,\,k} - \mu_{ik})(v_{j+1} - v_j), \qquad (4.49)$$

and

$$\log\theta_{ijk} = [\phi_{k+1}(\mu_{i+1,\,k+1} - \mu_{i,\,k+1}) - \phi_k(\mu_{i+1,\,k} - \mu_{ik})](v_{j+1} - v_j). \quad (4.50)$$

(c) Homogeneous μ_i and v_j, heterogeneous ϕ

$$\log F_{ijk} = \lambda + \lambda_i^A + \lambda_j^B + \lambda_k^C + \lambda_{ik}^{AC} + \lambda_{jk}^{BC} + \phi_k\mu_i v_j. \qquad (4.51)$$

The model has $(I - 1)(J - 1)K - K - (I - 2) - (J - 2) = IJK - (I + J - 1)$ $(K + 1) - 1$ df and the conditional log-odds ratios and the local log-odds ratios, respectively, are as follows:

$$\log\theta_{ij(k)} = \phi_k(\mu_{i+1} - \mu_i)(v_{j+1} - v_j), \qquad (4.52)$$

and

$$\log\theta_{ijk} = (\phi_{k+1} - \phi_k)(\mu_{i+1} - \mu_i)(v_{j+1} - v_j). \qquad (4.53)$$

This is a simple heterogeneous $RC(1)$ model and can be treated as a special case of the log-multiplicative layer effect model (LL_2) under Equation

4.10, except that the term ψ_{ij} is now replaced by μ_i and ν_j. Under Equation 4.51, it is easy to verify the following relationships:

$$\log\theta_{ij(k)} - \log\theta_{ij(k')} = (\phi_k - \phi_{k'})(\mu_{i+1} - \mu_i)(\nu_{j+1} - \nu_j). \qquad (4.54)$$

$$\frac{\log\theta_{ij(k)}}{\log\theta_{ij(k')}} = \frac{\phi_k}{\phi_{k'}}. \qquad (4.55)$$

$$\frac{\log\theta_{ij(k)} - \log\theta_{ij(k')}}{\log\theta_{ij(k)} - \log\theta_{ij(k*)}} = \frac{\phi_k - \phi_{k'}}{\phi_k - \phi_{k*}}. \qquad (4.56)$$

Equation 4.55 is equivalent to Equation 4.13 under the log-multiplicative layer effect model, and Equation 4.56 is equivalent to Equation 4.20 under the regression-type layer effect model. In other words, under the simple heterogeneous $RC(1)$ formulation, the ratios of the conditional log-odds ratio and the ratios of cross-layer differences in log-odds ratios are both proportional.

(d) Special cases

If the model under scenario (c) fits the data relatively well and the layer variable indexes temporal order, then models with linear or quadratic constraints listed under Equations 4.32 and 4.33 can be used. Similarly, if the layer variable indexes multiple groupings (say, ethnicity by gender), then the ANOVA-like constraints listed under Equations 4.36 to 4.38 can be applied as well.

However, the introduction of partial homogeneous or partial heterogeneous constraints to two-dimensional conditional association models, $RC(2)$-L, can be problematic even though they have been discussed at length elsewhere (Becker, 1989a, 1989b; Becker & Clogg, 1989). First, Wong (2001) claimed that some previously discussed partial homogeneous models have incorrect reported degrees of freedom because they either do not require or require fewer than expected cross-dimensional constraints.[5] Second, previous published examples are based on two groups. Since most social science applications naturally involve multiple group comparisons, it is unclear whether previous calculations can be generalized when K is greater than 2. As we shall find out shortly, some of them require only 1, not K, cross-dimensional constraint. Both issues are now addressed in Table 4.2.

Following the format similar to the previous table, Table 4.2 summarizes a series of partial homogeneous or partial heterogeneous $RC(2)$-L models. It should be noted that the proper accounting of the degrees of freedom of

Table 4.2 Cross-Dimensional Constraints and Degrees of Freedom for $RC(2)$-L Models With Partial Homogeneous or Partial Heterogeneous Constraints

Model	Cross-Dimensional Constraints	df
1. $\begin{bmatrix} 0 & 0 & 1 \\ 0 & 0 & 1 \end{bmatrix}$	$\sum \mu_{i11}\mu_{i21} = \sum \nu_{j1}\nu_{j2} = 0$ or $\sum \mu_{i1K}\mu_{i2K} = \sum \nu_{j1}\nu_{j2} = 0$	$(I-1)(J-1)K - 2(IK + J - K - 3)$
2. $\begin{bmatrix} 0 & 1 & 0 \\ 0 & 1 & 0 \end{bmatrix}$	$\sum \mu_{i1}\mu_{i2} = \sum \nu_{j11}\nu_{j21} = 0$ or $\sum \mu_{i1}\mu_{i2} = \sum \nu_{j1K}\nu_{j2K} = 0$	$(I-1)(J-1)K - 2(I + JK - K - 3)$
3. $\begin{bmatrix} 0 & 0 & 0 \\ 0 & 1 & 1 \end{bmatrix}$	None	$(I-1)(J-1)K - (I+J)(K+1) - 2(K+2)$
4. $\begin{bmatrix} 0 & 1 & 1 \\ 0 & 0 & 0 \end{bmatrix}$	None	$(I-1)(J-1)K - (I+J)(K+1) - 2(K+2)$
5. $\begin{bmatrix} 0 & 0 & 0 \\ 1 & 1 & 1 \end{bmatrix}$	None	$(I-1)(J-1)K - (K+1)(I+J-3)$

Model	Cross-Dimensional Constraints	df
6. $\begin{bmatrix} 1 & 1 & 1 \\ 0 & 0 & 0 \end{bmatrix}$	None	$(I-1)(J-1)K - (K+1)(I+J-3)$
7. $\begin{bmatrix} 0 & 0 & 1 \\ 0 & 1 & 1 \end{bmatrix}$	$\sum v_{j1}v_{j2} = 0$ or $\sum \mu_{i11}\mu_{i2} = 0$	$(I-1)(J-1)K - (I+IK+2J-7)$
8. $\begin{bmatrix} 0 & 1 & 0 \\ 0 & 1 & 1 \end{bmatrix}$	$\sum \mu_{i1}\mu_{i2} = 0$ or $\sum v_{j11}v_{j2} = 0$	$(I-1)(J-1)K - (2I+J+JK-7)$
9. $\begin{bmatrix} 0 & 1 & 0 \\ 1 & 1 & 1 \end{bmatrix}$	$\sum \mu_{i1}\mu_{i2} = 0$	$(I-1)(J-1)K - (2I+J+JK-K-6)$
10. $\begin{bmatrix} 0 & 0 & 1 \\ 1 & 1 & 1 \end{bmatrix}$	$\sum v_{j1}v_{j2} = 0$	$(I-1)(J-1)K - (I+IK+2J-K-6)$

(Continued)

Table 4.2 (Continued)

Model	Cross-Dimensional Constraints	df
11. $\begin{bmatrix} 0 & 1 & 1 \\ 0 & 1 & 1 \end{bmatrix}$	None	$(I-1)(J-1)K - 2(I+J+K-4)$
12. $\begin{bmatrix} 0 & 1 & 1 \\ 1 & 1 & 1 \end{bmatrix}$	None	$(I-1)(J-1)K - (2I+2J+K-7)$
13. $\begin{bmatrix} 1 & 1 & 1 \\ 0 & 1 & 1 \end{bmatrix}$	None	$(I-1)(J-1)K - (2I+2J+K-7)$

NOTE: Model entries of 0 mean no restrictions across groups, whereas entries of 1 mean equality restrictions across groups. See text for details.

any model, here and elsewhere, follows exactly the rules outlined earlier under the section on Model Estimation, Degrees of Freedom, and Model Selection in Chapter 2. To maintain maximum continuity and consistency, models presented in the first column follows the same nomenclature as in Becker and Clogg (1989), with the number of row entries representing dimensionality (i.e., $RC(2)$ model has 2 rows) and the number of column entries representing the number of interested parameters (3 columns for ϕ, μ_i, and v_j, respectively). Finally, all entries can have values of 0 or 1; 0 means no homogeneous equality constraints across layers, whereas 1 means the existence of such restrictions.

The first model imposes no equality constraints on the intrinsic association and row score parameters (i.e., ϕ_{mk} and μ_{imk}), but it imposes equality constraints on the column score parameters (v_{jm}) in both dimensions. Contrary to expectation, even though the cross-classified table involves K layers, the model requires only 2, not $K + 1$, cross-dimensional constraints. More specifically, one can either impose $\sum_{i=1}^{I} \mu_{i11}\mu_{i21} = \sum_{j=1}^{J} v_{j1}v_{j2} = 0$, $\sum_{i=1}^{I} \mu_{i1k}\mu_{i2k} = \sum_{j=1}^{J} v_{j1}v_{j2} = 0$, or cross-dimensional constraint with row scores at any layer together with another one with column scores. Note that the imposition of any unnecessary additional constraints would yield different test statistics and generally poorer fitting models than the one without constraints. As a result, the model has $(I-1)(J-1)K - 2(IK + J - K - 3)$ df. Similarly, when equality constraints are applied to the row score parameters at both dimensions but not to the other two parameters, only 2 but not $K + 1$ cross-dimensional constraints are needed. The latter model has $(I-1)(J-1)K - 2(I + JK - K - 3)$ df.

The next four models, Lines 3 to 6, offer another interesting property of some partial homogenous or heterogeneous $RC(2)$-L conditional association models. The specification under Model 3 imposes equality constraints only to the row and column score parameters in the second dimension (μ_{i2} and v_{j2}) but without additional restrictions to remaining parameters (ϕ_{1k}, ϕ_{2k}, μ_{i1k}, and v_{j1k}). Contrary to expectation, no cross-dimensional constraints are needed, and all parameters are rotationally unique. In reality, one can easily verify the present claim by using multiple random start values in 1_{EM} or any statistical software that has the capability for not imposing cross-dimensional constraints in iterative cycles. For models that do not require cross-dimensional constraints, they should all have identical converged maximum likelihood estimates and goodness-of-fit statistics (see the section on Identifying Restrictions and Degrees of Freedom in Chapter 3 for details). Similarly, the same rule applies to Model 4 when homogeneous equality constraints are applied only to the first dimensional row and column scores (μ_{i1} and v_{j1}) but without further restrictions on the remaining parameters

(ϕ_{1k}, ϕ_{2k}, μ_{i2k}, and ν_{j2k}). Models under Lines 5 and 6 further illustrate that no cross-dimensional constraint is needed when we expand equality constraints to their respective intrinsic association parameters as well; that is, ϕ_1, μ_{i1}, and ν_{j1} or ϕ_2, μ_{i2}, and ν_{j2}, respectively.

On the other hand, models under Lines 7 to 10 need only one cross-dimensional constraint. They all share the following common feature: homogeneous equality constraints are imposed to either row or column score parameters in both dimensions and then to the remaining row or column score parameters in one but not both dimensions. The cross-dimensional constraint pertains to the one when homogenous equality constraints in row or column scores from both dimensions have been imposed. For example, under either Model 7 or Model 10, only $\sum_{j=1}^{J} \nu_{j1}\nu_{j2} = 0$ is needed. In the case of Models 8 and 9, only $\sum_{i=1}^{I} \mu_{i1}\mu_{i2} = 0$ is needed instead. Their proper degrees of freedom are presented in the third column.

Model 11 has been discussed extensively by Wong (2001) and is a substantively interesting model. Because the model has wide social science applications, a more detailed discussion is in order. This is the simple heterogeneous $RC(M)$ model which postulates homogeneity in row and column score parameters from all dimensions (μ_{im} and ν_{jm}) but heterogeneity in intrinsic association parameters (ϕ_{mk}). When $M = 2$, the model can be formally written as follows:

$$\log F_{ijk} = \lambda + \lambda_i^A + \lambda_j^B + \lambda_k^C + \lambda_{ik}^{AC} + \lambda_{jk}^{BC} + \phi_{1k}\mu_{i1}\nu_{j1} + \phi_{2k}\mu_{i2}\nu_{j2}. \quad (4.57)$$

The model has $(I-1)(J-1)K - 2(I+J+K-4)$ df. According to Wong (2001), the model can be reparameterized as follows:

$$\log F_{ijk} = \lambda + \lambda_i^A + \lambda_j^B + \lambda_k^C + \lambda_{ik}^{AC} + \lambda_{jk}^{BC} + \phi_1\mu_{i1}\nu_{j1}\eta_{k1} + \phi_2\mu_{i2}\nu_{j2}\eta_{k2}, (4.58)$$

with $\sum_{k=1}^{K} \eta_{k1}^2 = \sum_{k=1}^{K} \eta_{k2}^2 = 1$. Written in such form, the model bears a strong resemblance of the log-trilinear function using the canonical or parallel factor (CP) decomposition method popular in the psychometric literature. It is well-known that under normal circumstances, solutions obtained from the CP decomposition method are unique and require no rotational restrictions (Carroll & Chang, 1970; Harshman, 1970; Kruskal, 1977).[6] The conditional log-odds ratios and the local log-odds ratios, respectively, are as follows:

$$\log \theta_{ij(k)} = \sum_{m=1}^{2} \phi_{mk} (\mu_{i+1,\,m} - \mu_{im})(\nu_{j+1,\,m} - \nu_{im}), \quad (4.59)$$

and

$$\log\theta_{ijk} = \sum_{m=1}^{2} \left(\phi_{m,\,k+1} - \phi_{mk}\right)\left(\mu_{i+1,\,m} - \mu_{im}\right)\left(\nu_{j+1,\,m} - \nu_{jm}\right). \quad (4.60)$$

The above model can be treated as special cases of the log-multiplicative layer effect model (LL_2) and regression-type layer effect model (LL_3) discussed in the previous section. It is a special case of LL_2 because although the postulated effects are more complex, layer differences nonetheless are still log-multiplicative in scale (ϕ_{1k} and ϕ_{2k}). On the other hand, it is a special case of LL_3 because λ_{ij}^{AB} and ψ_{ij} in Equation 4.16 are now replaced by (ϕ_{1k}, μ_{i1}, ν_{j1}) and (μ_{i2}, ν_{j2}), respectively. Unfortunately, under the current formulation, neither the ratios of the conditional log-odds ratio nor the ratios of cross-layer differences in log-odds ratios are proportional and can only be expressed in complex forms:

$$\frac{\log\theta_{ij(k)}}{\log\theta_{ij(k')}} = \frac{\sum_{m=1}^{2} \phi_{mk}\left(\mu_{i+1,\,m} - \mu_{im}\right)\left(\nu_{j+1,\,m} - \nu_{jm}\right)}{\sum_{m=1}^{2} \phi_{mk'}\left(\mu_{i+1,\,m} - \mu_{im}\right)\left(\nu_{j+1,\,m} - \nu_{jm}\right)}. \quad (4.61)$$

$$\frac{\log\theta_{ij(k)} - \log\theta_{ij(k')}}{\log\theta_{ij(k)} - \log\theta_{ij(k*)}} = \frac{\sum_{m=1}^{2} \left(\phi_{mk} - \phi_{mk'}\right)\left(\mu_{i+1,\,m} - \mu_{im}\right)\left(\nu_{j+1,\,m} - \nu_{jm}\right)}{\sum_{m=1}^{2} \left(\phi_{mk} - \phi_{mk*}\right)\left(\mu_{i+1,\,m} - \mu_{im}\right)\left(\nu_{j+1,\,m} - \nu_{jm}\right)}. \quad (4.62)$$

Note that when ϕ_{1k} is restricted as homogeneous as ϕ_1 in Equation 4.57, then the above $RC(2)$-L model with CP decomposition can be shown to share similar structure as Goodman and Hout's regression-type layer effect model and is termed as the *regression type RC effect model* (Yamaguchi, 1998, p. 241).

The last two models listed in Table 4.2 are based on the CP specification under Model 11, but with added equality constraints on one but not both intrinsic association parameters (i.e., either ϕ_{1k} or ϕ_{2k}). Because both models can still be represented in log-trilinear terms, no cross-dimensional constraints are needed, and the converged estimates are rotationally unique. The conditional log-odds ratios and the local log-odds ratios from Models 12 and 13 follow similar structures listed in Equations 4.54 and 4.55, except that the ϕ_{mk} parameters can be further simplified as either ϕ_1 or ϕ_2, depending on which equality constraints have been imposed.

A systematic comparison between various partial homogeneous or partial heterogeneous models listed above can help researchers locate specifically where homogeneity or heterogeneity constraints are needed. Further restrictions on ϕ_{mk} parameters are possible for models using the CP decomposition method, that is, Models 11 to 13, if the layer variable indexes temporal order, multiple groupings, or both. Furthermore, it should also be noted that when the conditional $RC(2)$-L models do not fit the data, it is possible to use hybrid models such as $U+RC(2)$-L, $R+RC(2)$-L, $C+RC(2)$-L, and $R+C+RC(2)$-L as well. In some occasions, the inclusion of nonparametric effects such as diagonal parameters in the case of squared tables may be more appealing. Finally, one can also consider another class of models with a priori fixed scores, $U_1^0, U_2^0, \ldots, U_M^0$, which is a generalization of the linear-by-linear association. The $U_M^0(m)-L$ or $\left(U_1^0 + U_2^0 + \cdots + U_M^0\right)-L$ model can be treated as a special case of the $(RC(M))$-L model. This is equivalent to the SAT model by Hout (1984, 1988) and has been used successfully to explore temporal changes in the American occupational mobility structure.

Example 4.1: Changes in Association Between Education and Occupation

The first illustrative example examines the question whether there are any systematic changes in the relationship between education and occupation in the United States over time as well as whether men and women enjoy similar relationship. The cross-tabulated frequencies displayed in Table 4.3 are derived from *1972–2006 Cumulative General Social Survey* (Davis et al., 2007). There are two time periods: 1975 to 1980 and 1985 to 1990. Individual education is measured by degree attained rather than years of schooling attended to match closely with employers' demands on workers' credentials. The measure has four outcomes: (1) college or more, (2) junior college, (3) high school, and (4) less than high school. Individual occupation has five categories: (1) upper nonmanual (UNM), (2) lower nonmanual (LNM), (3) upper manual (UM), (4) lower manual (LM), and (5) farm (F). The sample is restricted to white men and women aged between 25 and 39. The analysis below is therefore based on a $4 \times 5 \times 2 \times 2$ table (Table 4.3), with a total sample size of 4,078.

A close inspection of the table reveals that as the American economy rapidly transformed from industrial into postindustrial, service economy, more workers today are located in nonmanual positions while fewer are in the manual and farming occupations. The observation is particularly true among women. Also, despite the fact that some entries in the last column (Farm) have few or no observations, it is reassuring that their impacts on the estimated associated parameters are negligible. This confirms an earlier claim that the impact of zero or sparse cells has relatively little effect on the stability of association parameters as the latter are derived from an entire row and/or column rather than individual cells.

Table 4.3 Temporal Changes in Association Between Education and Occupation Among White Americans Over Time, 1975 to 1990

Education Degree	Occupation									
	White Men					White Women				
	UNM	LNM	UM	LM	F	UNM	LNM	UM	LM	F
(A) 1975–1980										
College or more	201	29	8	13	5	152	29	2	8	0
Junior college	18	6	3	6	0	17	12	0	3	0
High school	109	74	164	89	16	101	336	9	134	2
Less than high school	7	6	45	30	6	7	41	7	63	0
(B) 1985–1990										
College or more	247	58	20	23	2	288	51	1	17	3
Junior college	48	11	16	13	1	47	38	2	18	0
High school	157	68	178	116	27	165	321	27	168	1
Less than high school	7	7	50	42	5	12	25	5	29	6

NOTE: The five occupational categories are upper nonmanual (UNM), lower nonmanual (LNM), upper manual (UM), lower manual (LM), and farm (F). The analysis pertains to men and women aged between 25 and 39 years. The data are derived from *1972–2006 Cumulative General Social Survey*. The total sample size is 4,078.

Table 4.4 reports a series of homogeneous and heterogeneous association models to explore how the association between education and occupation may have changed over time and differed by gender. The first model is the baseline model that postulates constant association. It can also be termed as homogeneous $RC(0)$-L model and has 48 df and L^2 of 1,371, indicating clearly that constant association between tables is not a plausible scenario. Lines 2 and 3 report the results from the homogeneous and heterogeneous $RC(1)$-L models, respectively. Although the homogeneous $RC(1)$-L model offers a dramatic improvement over the baseline model, the goodness-of-fit statistic still indicates that the model is not consistent with the data. Its heterogeneous counterpart, on the other hand, provides a dramatic improvement (24 df and L^2 of 69), though it is still not acceptable by conventional standard. Note that the value of the BIC statistic is much more negative for the homogeneous model (-205) than its heterogeneous counterpart (-130). Although one might be tempted to choose the homogeneous formulation as the final model, this is not necessarily a sound decision because other competing models may provide better results as well as more negative BIC statistics.

The next two sets of models (Lines 4 to 7) increase the dimensionality of association from 1 to either 2 or 3. The heterogeneous two-dimensional association model (Line 5) provides the best result so far (8 df and L^2 of 6).

Table 4.4 Results of General $RC(M)$-L Association as Applied to Table 4.3

Model Description	df	L^2	BIC	Δ	P
1. $RC(0)$-L (homogeneous)	48	1370.93	971.89	23.75	.000
2. $RC(1)$-L (homogeneous)	42	143.83	−205.33	6.39	.000
3. $RC(1)$-L (heterogeneous)	24	69.06	−130.46	3.04	.000
4. $RC(2)$-L (homogeneous)	38	117.38	−198.52	5.24	.000
5. $RC(2)$-L (heterogeneous)	8	5.83	−60.67	0.44	.666
6. $RC(3)$-L (homogeneous)	36	113.18	−186.10	5.19	.000
7. $RC(3)$-L (heterogeneous)	0	0.00			
8. $U+RC$ (homogeneous)	41	124.75	−216.10	5.58	.000
9. $U+RC$ (heterogeneous)	22	27.47	−155.42	1.50	.194

The contrast between homogeneous and heterogeneous $RC(2)$-L models reveals that with a difference of 30 df, there is a total variation of 111.6 (= 117.4 − 5.8) chi-square points, clearly indicating that the association between education and occupation is not the same across tables. While the heterogeneous $RC(3)$ model is a saturated model with 0 df, its homogeneous counterpart has 36 df and L^2 of 113. The last two models (Lines 8 and 9) are hybrid models using a slightly different two-dimensional formulation (U+RC) to capture the complex association structure. In particular, the goodness-of-fit statistic from the heterogeneous U+RC model also yields satisfactory result (20 df, L^2 of 27, and p = .12). The contrast between both heterogeneous and homogeneous models also informs significant variation across tables (close to 97 chi-square points with 16 df). While it is possible to use other hybrid formulations such as R+RC, C+RC, and R+C+RC to better understand the relationship involved, the discussion below focuses on the $RC(2)$-L models instead to further illustrate situations where cross-dimensional constraints can be relaxed or reduced and the proper accounting of the degrees of freedom can be done if needed.[7]

Using models listed in Table 4.4, it is possible to decompose their goodness-of-fit statistics to locate the exact dimension of association deemed important to understand the relationship involved. Two partitioning results are reported in Table 4.5. In Panel A, the partition exercise is based on various complete homogeneous models. For example, the contrast between Models 1 and 2 reveals that a substantial majority of the association between education and occupation lies in the first dimension (89.5% with 6 df), leaving 2% and 0.3% to second and third dimension, respectively, and the remaining 8.3% are due to heterogeneity in the log-multiplicative components. Similar results based on the complete heterogeneous models can be found under panel B. While a substantial majority of association lies in the first dimension (95%), the contrast between Models 3 and 5 informs that slightly more than 4.6% of the total variation is due to the second dimension, leaving less than 0.5% to the third dimension. In other words, results from both panels confirm that the $RC(2)$-L formulation best captures the overall association between education and occupation, and there are substantial differences across tables.

Using the heterogeneous $RC(2)$ model as the starting point, a series of models with partial homogeneity constraints are reported in Table 4.6. With six sets of parameters under consideration (ϕ_{1k}, ϕ_{2k}, μ_{i1k}, μ_{i2k}, ν_{j1k}, and ν_{j2k}), we need to develop a sound and systematic strategy. As a simple rule of thumb, tests for homogeneous equality constraints are first applied to the row and/or column score parameters. When their results are supportive, further homogeneous equality constraints on the intrinsic association parameters are then imposed. This strategy, when applied appropriately, is

Table 4.5 Analysis of Association for *RC(M)-L* Models as Applied to
Table 4.4

Source	Models Used	L^2 Component	Percentage of Total	Degrees of Freedom
(A) Components based on complete homogeneous models				
First dimension	(1) − (2)	1227.10	89.50	6
Second dimension	(2) − (4)	26.45	1.93	4
Third dimension	(4) − (6)	4.20	0.31	2
Heterogeneity	(6)	113.18	8.26	36
Total	(1)	1370.93	100.00	48
(B) Components based on complete heterogeneous models				
First dimension	(1) − (3)	1301.87	94.96	24
Second dimension	(3) − (5)	63.23	4.61	16
Third dimension	(5) − (7)	5.83	0.43	8
Total	(1)	1370.93	100.00	48

likely to provide simple and substantively interpretable results. Note that
there are probably more reported models than necessary in Table 4.6. Their
inclusion here is to verify claims from the previous section that a number
of them require fewer or even no cross-dimensional constraints.

The first four models provide different combinations to test whether row
and column scores from both dimensions are equal across gender and over
time (recall that the layer variable is a 2×2 grouping variable). Each represents different tests: Model 1 imposes equal row scores in both dimensions;
Model 2 imposes equal column scores in both dimensions; Model 3 imposes
equal row and column scores in the first dimension; and finally, Model 4
imposes equal row and column scores in the second dimension. With the
exception of the first one, there is strong evidence to suggest that row and
column scores in both dimensions are indeed equal across layers. For example, the model with equal column scores in both dimensions have 14 *df* and
L^2 of 8, indicating that there is very little empirical evidence to reject the null
hypothesis that the row score parameters are equal across layers. According

Table 4.6 Results of Partial Homogeneous or Partial Heterogeneous $RC(2)$-L Models

Model Description	df	L^2	BIC	Δ	P
1. Homogeneous μ_{i1} and μ_{i2}	14	7.98	−108.42	0.69	.891
2. Homogeneous ν_{j1} and ν_{j2}	20	29.25	−137.02	1.82	.083
3. Homogeneous μ_{i1} and ν_{j1}	15	11.21	−113.49	0.89	.737
4. Homogeneous μ_{i2} and ν_{j2}	15	16.90	−107.80	1.46	.325
5. Homogeneous μ_{i1}, μ_{i2}, and ν_{j1}	22	19.61	−163.28	1.23	.608
6. Homogeneous μ_{i1}, ν_{j1}, and ν_{j2}	25	33.01	−174.82	2.10	.130
7. Homogeneous ϕ_1, μ_{i1}, and ν_{j1}	18	12.35	−137.29	1.00	.829
8. Homogeneous ϕ_2, μ_{i2}, and ν_{j2}	18	18.21	−131.43	1.24	.442
9. Homogeneous μ_{i1}, μ_{i2}, ν_{j1} and ν_{j2}	30	38.46	−210.94	2.12	.138
10. Homogeneous μ_{i1}, μ_{i2}, ν_{j1}, and ν_{j2}, restricted ϕ_1 (constant by gender over time)	32	40.44	−225.59	2.17	.146
11. Homogeneous ϕ_1, μ_{i1}, μ_{i2}, ν_{j1}, and ν_{j2}	33	43.44	−230.90	2.31	.106
12. Homogeneous ϕ_1, μ_{i1}, μ_{i2}, ν_{j1}, and ν_{j2}, restricted ϕ_2 (same for women over time)	34	44.64	−238.01	2.70	.105
13. Homogeneous ϕ_1, ϕ_2, μ_{i1}, μ_{i2}, ν_{j1}, and ν_{j2}	38	117.38	−198.52	5.24	.000

to Table 4.2, Models 1 and 2 require two cross-dimensional constraints, whereas Models 3 and 4 require no restrictions at all.

The next two models attempt to test equality constraints in three terms: Model 5 (μ_{i1}, μ_{i2}, and ν_{j1}) and Model 6 (μ_{i1}, ν_{j1}, and ν_{j2}). Both models require only one cross-dimensional constraint. Based on the goodness-of-fit statistics, the restrictions appear to be more appropriate in the former (22 df and $L^2 = 20$) than the latter (25 df and $L^2 = 33$). Model 7 postulates homogeneity of all parameters in the first dimension (ϕ_1, μ_{i1}, and μ_{j1}), whereas Model 8 postulates homogeneity of all parameters in the second dimension (ϕ_2, μ_{i2}, and v_{j2}). No cross-dimensional constraint is needed in either case, and the results are both satisfactory (L^2 of 12 and 18, respectively, with 18 df).

Model 9 adopts the CP decomposition method and postulates heterogeneity in the two intrinsic association parameters, ϕ_{1k} and ϕ_{2k}, instead. The goodness-of-fit of the model is moderately good (30 df and $L^2 = 38$). Relative to all other models discussed so far, Model 9 would have been preferred because it has the most negative BIC value among all $RC(M)$ models. Furthermore, a comparison of the present model with its complete homogeneous and heterogeneous counterparts (Models 4 and 5 in Table 4.4) informs that about 78.9 out of 111.6 chi-square points or 70.7% of the total variation can be captured by the eight intrinsic (ϕ_{1k} and ϕ_{2k}) parameters, whereas cross-layer variation in the row and column scores (μ_{i1k}, μ_{i2k}, v_{j1k}, and v_{j2k}) accounts for the remaining 29.3%.

Conditional on Model 9, the last four models attempt to impose further restrictions on the intrinsic association parameters in both dimensions. A close inspection of ϕ_{1k} suggests that there is little or no variation by gender over time in the first dimension, and this hypothesis is tested in Model 10. Relative to the previous model, a gain of two additional degrees of freedom does not lead to any significant deterioration in fit. Further inspection reveals that the intrinsic association parameters for men and women in the first dimension now appear to be invariant as well. This hypothesis is then tested in Model 11, and the relative fit of the latter model is satisfactory. Based on the estimates from this model, it reveals further that while the intrinsic association parameters for women in the second dimension remain the same, those pertaining to men do not. The hypothesis with this particular restriction is tested in Model 12 and cannot be rejected.

Finally, the complete homogeneous $RC(2)$ model from Table 4.4 is reported again as Model 13. By now, it is obvious that restricted changes in ϕ_{1k} and ϕ_{2k} offer the most parsimonious and powerful way to understand temporal changes and gender differences in association between education and occupation in America. By using only four parameters in ϕ_{1k} and ϕ_{2k}, the preferred model (Model 12) captures about 65% of the total variation. Furthermore, the preferred model not only achieves acceptance by conventional standard but also has the most negative value in the BIC statistic. In other words, the so-called trade-off between model accuracy and scientific parsimony does not seem to be real when alternative competing models are involved.

The parameter estimates of Models 9 and 12 are reported in Table 4.7, together with their asymptotic standard errors (obtained directly from the *gnm* module in *R*). As expected, the estimated row and column score parameters from both models are highly correlated and exhibit similar patterns. Using symmetrical normalization (Equation 2.38) on the parameter estimates from

Table 4.7 Parameter Estimates of Changes in Association Between Education and Occupation Over Time, 1975 to 1990

Parameters		Model 9		Model 12	
		1st Dim	2nd Dim	1st Dim	2nd Dim
ϕ_{mk}	Men, 75–80	3.075	0.539	2.979	0.731
		(0.399)	(0.443)	(0.336)	(0.467)
	Women, 75–80	3.474	1.686	2.979	1.973
		(0.537)	(1.124)	(0.336)	(0.972)
	Men, 85–90	2.949	−0.770	2.979	−0.705
		(0.411)	(0.484)	(0.336)	(0.421)
	Women, 85–90	2.460	1.892	2.979	1.973
		(0.382)	(1.110)	(0.336)	(0.972)
μ_{im}	College+	−0.640	0.731	−0.650	0.770
		(0.029)	(0.079)	(0.027)	(0.060)
	Junior college	−0.239	−0.217	−0.227	−0.165
		(0.039)	(0.141)	(0.039)	(0.143)
	High school	0.168	−0.636	0.171	−0.617
		(0.031)	(0.083)	(0.028)	(0.083)
	<High school	0.711	0.121	0.705	0.012
		(0.021)	(0.171)	(0.019)	(0.167)
ν_{jm}	Upper nonmanual	−0.765	0.071	−0.748	0.043
		(0.025)	(0.101)	(0.030)	(0.077)
	Lower nonmanual	−0.250	−0.480	−0.280	−0.477
		(0.046)	(0.073)	(0.050)	(0.073)
	Upper manual	0.398	−0.198	0.393	−0.227
		(0.052)	(0.105)	(0.053)	(0.102)
	Lower manual	0.273	−0.216	0.259	−0.169
		(0.048)	(0.066)	(0.047)	(0.063)
	Farm	0.344	0.824	0.376	0.831
		(0.086)	(0.039)	(0.086)	(0.031)

NOTE: Values in parentheses are the asymptotic standard errors.

Model 9, Figures 4.1 and 4.2 provide graphical representations of the estimated row scores (education) and column scores (occupation), respectively. In terms of education, it is obvious that the first dimension is vertical (i.e., higher levels of educational attainment have higher scores), whereas the second dimension represents departures from that vertical image. Similarly, the estimated column scores (occupation) in the first dimension conform largely to conventional socioeconomic rankings: nonmanual occupations rank higher than manual and farm occupations. The only complication is that lower manual tends to have higher relative rankings than farm and upper manual. The distinction between them, however, is small in the first dimension. Rather, most of the deviations occur in the second dimension between manual and farm occupations. This anomaly may be in part due to sample restriction, as the analysis pertains only to individuals who are still at their early or mid-careers (aged between 25 and 39 years). Nevertheless, the row and column scores in the first dimension together reflect socioeconomic hierarchy or vertical image in status attainment, that is, the higher an individual's educational attainment, the higher is his or her socioeconomic attainment.

According to Model 12, there is no gender disparity and temporal change for the intrinsic association parameter in the first dimension.[8] All gender and temporal differences lie in the second dimension instead. If the first dimension denotes socioeconomic hierarchy, then the second dimension can be interpreted as additional channels and/or barriers that deviate from that hierarchy. Given that the intrinsic association parameters have remained strong and stable among American women ($\phi_2 = 1.973$) for both periods, these additional channels and barriers have persisted over time. Perhaps, the most interesting observation is that while the channels and barriers were applicable to American men in 1975 to 1980 as well, albeit at a much weaker level than their female counterparts, these effects were reversed in 1985 to 1990.

What are these additional channels and/or barriers then? First, everything else being equal, individuals with junior college and high school education are more likely to enter lower nonmanual occupations (recall that the product of row and column scores in the second dimension is positive). This is consistent with the feminization of nonmanual labor, especially in clerical, secretarial, and sales positions that is also common in other advanced industrial societies. Second, there is a surprisingly larger than expected propensities for college educated individuals to work in farm occupations. Of course, this does not necessarily represent a significant development given that the actual number of people involved is rather small for both sexes. Third, college-educated individuals are relatively less likely to wind up in lower nonmanual and manual occupations. This probably reflects the general trend toward professionalization in many nonmanual positions. While the sources

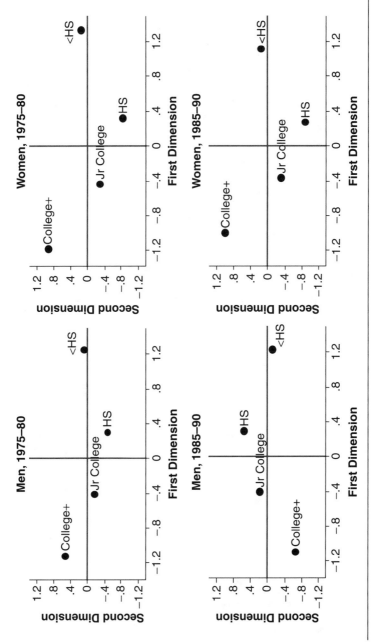

Figure 4.1 Estimated Education Scores From Model 9 of Table 4.6

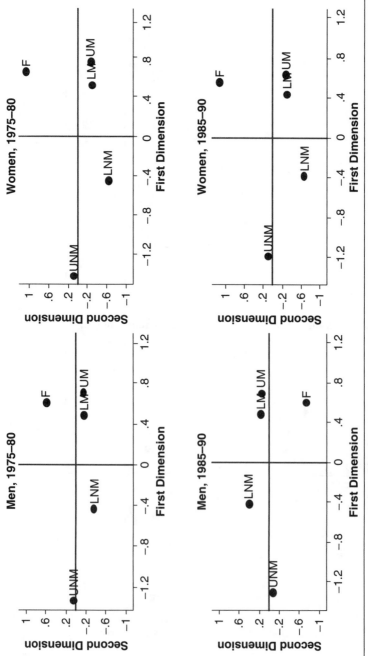

Figure 4.2 Estimated Occupation Scores From Model 9 of Table 4.6

of such additional channels and barriers may be driven by both labor market demands and supply due to individual tastes and preferences, they nonetheless represent a rather clear-gendered social division of labor.

Example 4.2: Relationship Between Education and Attitude Toward Premarital Sex

The second example examines possible temporal change in the relationship between education and attitude toward premarital sex in the United States. Entries in Table 4.8 are tabulated from the *1972–2006 Cumulative General Social Survey* (Davis et al., 2007) and are restricted to women only. There are altogether 21 tables, and the total sample size is 16,548. Education is measured by four categories: (1) less than high school, (2) high school, (3) some college, and (4) college or more. Attitude toward premarital sex has four outcomes: (1) always wrong, (2) almost always wrong, (3) sometimes wrong, and (4) not wrong. The analysis below therefore pertains to a $4 \times 4 \times 21$ table. The working hypothesis here is that if there is a secular trend toward individual freedom and women's greater control of their own bodies and that better educated women are more likely to subscribe to such ideology, then the strong association between (lower) education and (disapproval of) premarital sex should decline over time.

A series of models is presented in Table 4.9 to detect possible changes over time. The conditional independence model (Line 1) is used as the baseline model for comparison. It has 189 df and L^2 of 651, with close to 8% of women misclassified under such formulation. On the other hand, the full two-way interaction or the homogeneous association model provides a dramatic improvement (180 df and L^2 of 193), representing more than 70% reduction in the goodness-of-fit statistic. While it is tempting to choose the homogeneous association as reasonable approximation, it should be checked against other models that examine temporal change explicitly. Two layer effects, LL_1 and LL_2, are estimated and reported in Lines 3 and 4. They yield very similar results though the log-multiplicative specification is slightly more preferred. However, with 20 df and ΔL^2 of 19 or 22 for LL_1 and LL_2, respectively, there is little evidence to suggest any temporal trend in the (unstructured) odds ratios. At the same time, part of the reason for its failure to detect change can be the lack of the statistical power in the test itself. Instead one should use the more powerful 1-df parametric test to detect visible trend, if any. When the linear trend restriction (Line 5) has been imposed, it is only slightly more preferred than its homogeneous counterpart. Given that their values in the *BIC* statistics are virtually identical, the model with no change should be preferred instead (Raftery, 1996; Wong, 1994). In other words, a significant chi-square statistic of 5.25 with 1 df may simply be the outcome of large sample size.

Table 4.8 Cross-Classification of Education and Attitude Toward Premarital Sex Among Women

Education	1972				1974				1975			
	1	2	3	4	1	2	3	4	1	2	3	4
<HS	170	29	59	50	129	32	46	33	124	42	63	60
HS	105	46	76	50	88	48	85	91	99	42	58	85
Some college	43	9	27	28	38	11	35	38	32	15	30	45
College+	13	9	23	21	20	13	25	34	17	12	38	23

Attitude Toward Premarital Sex

Education	1977				1978				1982			
	1	2	3	4	1	2	3	4	1	2	3	4
<HS	139	28	65	59	125	31	41	91	164	37	58	91
HS	98	32	68	106	106	38	78	103	112	30	82	135
Some college	33	15	36	40	45	18	32	61	49	17	41	94
College+	28	10	12	37	17	13	27	40	25	16	26	63

Attitude Toward Premarital Sex

Education	1983				1985				1986			
	1	2	3	4	1	2	3	4	1	2	3	4
<HS	105	27	38	76	102	19	29	70	100	26	40	70
HS	96	43	91	98	103	28	61	99	98	33	74	94
Some college	38	18	47	64	43	14	46	88	44	5	44	61
College+	38	14	37	59	27	8	21	57	30	8	31	58
	1988				1989				1990			
	1	2	3	4	1	2	3	4	1	2	3	4
<HS	66	19	25	37	59	12	23	42	43	14	20	30
HS	40	22	43	63	75	19	38	63	47	21	35	65
Some college	25	17	30	52	23	17	35	59	31	18	30	44
College+	20	9	29	35	21	5	26	37	25	12	23	42
	1991				1993				1994			
	1	2	3	4	1	2	3	4	1	2	3	4
<HS	64	14	21	43	47	17	20	33	88	21	32	61
HS	47	29	38	71	59	20	45	74	117	39	69	136

(Continued)

111

Table 4.8 (Continued)

Education	Attitude Toward Premarital Sex											
	1	2	3	4	1	2	3	4	1	2	3	4
Some college	39	12	25	61	34	19	28	66	78	35	54	115
College+	26	15	27	48	29	10	31	51	65	23	71	120
	1996				1998				2000			
<HS	68	21	31	66	60	23	35	45	70	18	29	67
HS	97	35	72	120	103	28	73	110	96	29	65	107
Some college	61	46	66	122	78	31	55	115	79	26	72	122
College+	58	21	68	110	59	26	58	113	62	21	54	96
	2002				2004				2006			
<HS	36	7	10	20	25	7	3	20	57	18	33	54
HS	57	14	34	59	54	12	22	46	90	23	61	105
Some college	40	15	25	64	48	16	33	63	107	30	60	146
College+	29	9	24	52	21	11	29	56	71	25	70	36

NOTE: The categories for premarital sex are always wrong (1), almost always wrong (2), sometimes wrong (3), and not wrong (4). The total sample size is 16,548.

Table 4.9 Results of Association Models as Applied to Table 4.8

Model Description	df	L^2	BIC	Δ	p
1. Conditional independence	189	650.81	−1185.14	7.84	.000
2. Homogeneous association	180	193.07	−1550.45	3.92	.169
3. Log-linear layer effect (LL_1)	160	174.04	−1380.21	3.64	.212
4. Log-multiplicative layer effect (LL_2)	160	170.83	−1383.41	3.59	.265
5. Model (4) with linear trend restriction	179	187.82	−1550.99	3.82	.311
6. Heterogeneous U	168	232.60	−1399.35	4.39	.001
7. Heterogeneous RC	84	94.45	−721.53	2.48	.204
8. Simple heterogeneous RC	164	178.42	−1414.68	3.75	.209
9. Model (8) with linear trend in ϕ	183	195.46	−1582.20	3.95	.251
10. Model (8) with quadratic trend in ϕ	182	195.36	−1572.59	3.96	.236
11. Homogeneous RC	184	206.45	−1580.93	4.08	.123
12. Model (8) + $v_3 = v_4$	165	179.55	−1423.27	3.79	.208
13. Model (12) with linear trend in ϕ	184	196.79	−1590.59	3.96	.246

Two special association models have been estimated to capture the relationship between education and attitude toward premarital sex. The first one is the heterogeneous uniform association (U) model (Line 6), and the second one is the heterogeneous log-multiplicative row and column effects (RC) model (Line 7). Result from the latter model is clearly satisfactory (84 df, L^2 of 94.5, and $p = .2$), but it is not the case for the former (168 df and L^2 of 233). The remaining models in the table impose various kinds of restrictions to the $RC(1)$ model to detect trend. The simple heterogeneous model in Line 8 is equivalent to the partial homogeneous RC model on row and column score parameters (μ_i and v_j), with only ϕ_k to capture temporal trend. The overall fit of this model is acceptable (164 df and L^2 of 178.4). Results from both linear trend (Line 9) and quadratic trend (Line 10) restrictions are virtually identical, indicating that the latter specification is clearly not consistent with the data.

Together with the homogeneous RC model in Line 11, one can determine the extent of changes in intrinsic association over the past 34 years. A contrast between Models 8 and 11 reveals that the total variation in ϕ_k is about 28 chi-square points with 20 df. The linear trend restriction alone with 1 df captures about 17 chi-square points or about 61% of that variation. Furthermore, the nested chi-square difference test between Models 9 and 11 is statistically significant at the 0.001 level, clearly rejecting any claim that the linear trend is the result of random noise or error. Note that the BIC statistic from Model 10 has by far the most negative value ($-1,582$) and is clearly more preferred over other simpler but probably incorrect models.

It is possible to arrive at even simpler models. A close examination of the parameter estimates from Model 8 reveals that the estimated column scores for attitudes toward premarital sex are extremely close for the "sometimes wrong" and "not wrong" categories (i.e., $v_3 = v_4$). When such constraint has been imposed alone (Line 12) and together with linear trend restriction (Line 13), they both yield acceptable results and should be preferred. In sum, both Models 9 and 13 can be taken to understand subtle changes in the association between education and attitudes toward premarital sex over time. The example reiterates again that the trade-offs between model accuracy and scientific parsimony may have been overstated or abused by some empirical practitioners since it is possible to achieve both goals simultaneously.

The parameter estimates and their asymptotic standard errors from Models 9 and 13 are presented in Table 4.10.[9] With the exception of v_3 and v_4, there is virtually no difference between the parameter estimates reported for both models. Generally speaking, less educated women expressed much more conservative views toward premarital sex, disapproving it as "always wrong," whereas college educated women are more likely to tolerate or condone such behavior. They tend to regard premarital sex as either "sometimes wrong" or "not wrong at all." In other words, the results here confirm that there is a strong relationship between education and attitude toward premarital sex. Such relationship, however, is not static, and its strength has experienced subtle but gradual decline between 1972 and 2006 (see also Figure 4.3). While the decline in intrinsic association (ϕ) is relatively small on an annual basis (0.010), a cumulative drop of more than 40% from 0.812 to 0.482 is nonetheless substantial. However, without the use of these statistically powerful tests, it is unlikely that such subtle trend would have been detected. In sum, the utility of various conditional association models has been successfully illustrated from the two examples. In both cases, the final models offer simple and substantively interpretable results. Perhaps more significantly, our choice of the final model satisfies two seemingly contradicting demands—model accuracy and scientific parsimony.

Table 4.10 Parameter Estimates of Temporal Changes in Association Between Education and Attitude Toward Premarital Sex Among American Women, 1972 to 2006

Parameters		Model 9	Model 13
ϕ_t (baseline)		0.812	0.811
		(0.068)	(0.066)
Linear trend in ϕ_t		−0.010	−0.010
		(0.003)	(0.003)
μ_i	<HS	−0.769	−0.771
		(0.017)	(0.017)
	HS	−0.086	−0.082
		(0.035)	(0.035)
	Some college	0.296	0.292
		(0.041)	(0.042)
	College+	0.560	0.561
		(0.035)	(0.035)
ν_j	Always wrong	−0.771	−0.771
		(0.028)	(0.028)
	Almost always wrong	−0.111	−0.115
		(0.061)	(0.061)
	Sometimes wrong	0.398	0.443
		(0.043)	(0.016)
	Not wrong	0.484	0.443
		(0.038)	(0.016)

NOTE: Values in parentheses are the asymptotic standard errors.

116

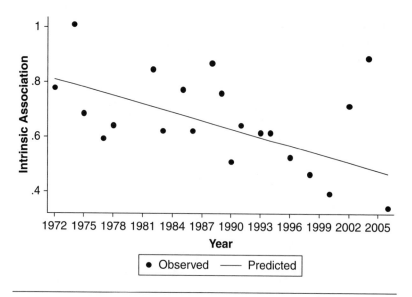

Figure 4.3   Temporal Changes in Association Between Education and
Premarital Sex

Notes

1. A distribution is defined as isotropic if the rows and columns can be ordered in such a way that the odds ratios $\theta_{ij} \geq 1$ for all i and j (Yule, 1906; see also Goodman, 1981b, for further discussion).

2. Note that the ratios of cross-layer differences of Equation 4.20 would also be applicable for both log-linear and log-multiplicative layer effect models as well. For example, the cross-layer ratios for LL_1 and LL_2 are $[(\beta_k - \beta_{k'})]/[(\beta_k - \beta_{k^*})]$ and $[(\phi_k - \phi_{k'})]/[(\phi_k - \phi_{k^*})]$, respectively.

3. See Hout (1984) and Ishii-Kuntz (1991) for the application of the uniform association models to multiple tables.

4. This is not to be confused with equality constraints within the same set of parameters such as $\mu_3 = \mu_4$ and $\nu_2 = \nu_5$. The latter constraints will use the term *equal* instead of *homogeneous*.

5. The author thanks Jeroen Vermunt, via personal communication, for helpful insights about the potential pitfalls and solutions of cross-dimensional constraints for these types of models.

6. The practical use of the *CP* decomposition method is sometimes complicated by the occurrence of the so-called degenerate sequence of *CP* solutions. When they occur, the convergence of the *CP* algorithm becomes extremely slow (say, more than 200,000 iterations in 1_{EM}) and some components of the *CP* solutions become more

and more correlated as the *CP* algorithm runs (Stegeman, 2007). The occurrence of degenerate sequence of *CP* solutions is due to the fact that the *CP* objective function has no minimum, but rather an infinum, and therefore the sequence of *CP* solutions fail to converge and display the pattern of degeneracy (Kruskal, Harshman, & Lundy, 1989). Degenerate sequence of *CP* solutions can be avoided by imposing orthogonality constraints on the component matrices (Harshman & Lundy, 1984). This, of course, will result in some loss of fit, and the degrees of freedom of the model would need to be adjusted accordingly (Stegeman, 2007, p. 603).

7. On the other hand, these hybrid models may be attractive among some social science practitioners because they do not need to consider cross-dimensional constraints and the proper accounting of the degrees of freedom of a particular model.

8. According to Yamaguchi (1998, p. 241), the conditional log-odds ratios under Model 12 can be expressed as Goodman and Hout's regression-type layer effect model.

9. The asymptotic, jackknife, and bootstrap standard errors are all extremely close, down to the third significant digits, for both highly parameterized models.

CHAPTER 5. PRACTICAL APPLICATIONS
OF ASSOCIATION MODELS

The utility of association models has been successfully illustrated in previous chapters to analyze two-way, three-way, and multiway tables. Not only do they provide insightful understanding of the underlying complex association patterns, the results also reveal that when the postulated (association) model is consistent with data, the use of conventional statistical testing as the criterion for choosing competing models (i.e., L^2 relative to its degrees of freedom) is still relevant. Also, the selection of the final preferred model seems to be unaffected by sample size, as the latter can vary from less than 1,000 to more than 16,000 in various illustrative examples. On the other hand, the influence of sample size appears to be problematic only when the specified model is "wrong." In other words, much of the discussion on the influence of sample size is predicated on wrong or improper model specification (Wong, 2003a). Under these circumstances, any small departures from the "true" model will be magnified by large samples. Without careful consideration of competing alternative models, any indiscriminate use of popular model selection criterion such as the *BIC* statistic will most likely lead to the acceptance of "wrong" models (Weakliem, 1999).

To further illustrate the potential usefulness of association models, the present chapter offers two practical applications. The first one deals with the question whether certain categories of the row and/or column variables can be combined. This nontrivial problem has been addressed thoroughly by Goodman (1981c), but the current discussion is now extended to include *RC(M)* association models to provide additional insights on the same subject matter. The findings below further reveal that researchers should also pay close attention to the other side of the aggregation problem and consider carefully the potential problems associated with the consequences of *inappropriate* aggregation. When applied incorrectly, researchers may unknowingly introduce systematic distortions in the underlying association patterns (see Wong, 2003b, for a thorough methodological critique in the case of marriage homogamy research). The second example addresses the potential use of *RC* association models as optimal scaling tools (see Rosmalen, Koning, & Groenen, 2009, for related discussion). Similar issues have been addressed before by using *RC* association models from two-way tables (Kateri & Iliopoulos, 2004; Smith & Garnier, 1987). It is now extended to include three-way tables (and possibly higher-order tables as well) via partial log-multiplicative association models to gain more reliable and valid scales that can be defended on both methodological and theoretical grounds. The construction of "proper" scales can then be used in subsequent multivariate analyses.

Example 5.1: Association Models to Determine Whether Some Categories Can Be Combined

The first example (Table 5.1) is taken from Guttman (1971), which crosstabulates 1,554 Israeli adults according to their "principal worries" and their living places as well as, in some cases, that of their fathers. The author would like to thank Ronald Breiger for bringing this particular example to his attention. It has been analyzed quite extensively via correspondence analysis (Greenacre, 1988). The row variable measures the principal worries (WOR) of an individual and has eight outcomes: (1) political situation (POL), (2) military situation (MIL), (3) economic situation (ECO), (4) enlisted relative (ENR), (5) sabotage (SAB), (6) more than one worry (MTO), (7) personal economics (PER), and (8) other worries (OTH). The

Table 5.1 Principal Worries and Living Situations Among Israeli Adults

Principal Worries	Living Situations				
	EUAM	IFEA	ASAF	IFAA	IFI
Political situation (POL)	118	28	32	6	7
Military situation (MIL)	218	28	97	12	14
Economic situation (ECO)	11	2	4	1	1
Enlisted relative (ENR)	104	22	61	8	5
Sabotage (SAB)	117	24	70	9	7
More than one worry (MTO)	42	6	20	2	0
Personal economics (PER)	48	16	104	14	9
Other (OTH)	128	52	81	14	12

SOURCE: Guttman (1971).

NOTE: The acronyms for living situations are EUAM, living in Europe or America; IFEA, living in Israel, father living in Europe or America; ASAF, living in Asia or Africa; IFAA, living in Israel, father living in Asia or Africa; and IFI, living in Israel and father also living in Israel. The total sample size is 1,554.

column variable indexes individual living situation and has five outcomes: (1) living in Europe or America (EUAM); (2) living in Israel, father living in Europe or America (IFEA); (3) living in Asia or Africa (ASAF); (4) living in Israel, father living in Asia or Africa (IFAA); and (5) living in Israel, father also living in Israel (IFI).

In addition to correspondence analysis, the table can be analyzed by canonical correlations (Gilula, 1986; Gilula & Haberman, 1988) and Goodman's $RC(M)$ association models. Because there is a close relationship between association model, correspondence analysis, and canonical correlation (Goodman, 1985; Greenacre, 1984), differences between these three types of analyses may not be large. Our underlying goal here is to determine whether some of the row and/or column categories can be combined that would result in even simpler understanding of the underlying association involved. Stated differently, we are interested to determine whether some rows and/or columns share similar propensities with each other. Of course, the question whether and which row and/or column categories can be combined is an empirical one, especially when the categories do not easily lend to simple rank ordering arrangements as in the current example.

Panel A of Table 5.2 provides a series of statistical models to understand the association between principal worries and living situations. The independence model has 28 df, L^2 of 121.5, with almost 10% of respondents misclassified under the model. On the other hand, the first dimensional RC association model, $RC(1)$, offers a dramatic improvement. The model has 18 df, L^2 of 29.2, indicating about 76% reduction in the likelihood ratio chi-square statistic and is marginally significant at the 0.05 level. Because the model is close to statistical acceptance, one may be tempted to impose equality constraints among row and column scores to achieve more parsimonious results while achieving statistical insignificance.

This can be achieved from models listed under Lines 3 and 4: Model 3 imposes the following equality restrictions on estimated row scores (MIL = ECO = MTO and ENR = SAR = OTH) and estimated column scores (ASAF = IFAA), whereas Model 4 imposes a further restriction that the above row scores are all equal (i.e., MIL = ECO = MTO = ENR = SAR = OTH). Relative to Model 2, a gain of 5 df in Model 3 does not lead to any significant deterioration of goodness-of-fit statistic. However, the further equality restriction between all six row scores can be rejected (1 df, L^2 of 4.9). Everything else being equal, Model 3 appears to be a perfect one because it is not statistically significant ($p = .16$) and has large negative BIC statistic. Such conclusion, however, may not be warranted.

Model 5 increases the dimensionality of the log-multiplicative RC association to two, and the result is highly satisfactory (10 df, L^2 of 6.8, $p = .74$).

Table 5.2 Analysis of Principal Worries Example

Model Description

A. Original table

	df	L^2	BIC	Δ	p
1. Independence model	28	121.47	−84.29	9.84	.000
2. RC(1) model	18	29.19	−103.08	4.01	.046
3. RC(1) model with equality restrictions MIL = ECO = MTO, ENR = SAB = OTH, and ASAF = IFAA	23	29.61	−139.41	4.08	.161
4. RC(1) model with equality restrictions MIL = ECO = MTO = ENR = SAB = OTH and ASAF=IFAA	24	34.51	−141.85	4.80	.076
5. RC(2) model	10	6.81	−66.68	0.87	.743
6. RC(2) model with equality restrictions ECO = MTO = MIL = ENR = SAB and ASAF = IFAA in both dimensions	20	13.66	−133.31	2.82	.847

Model Description

B. Analysis of association

	Contrast	L^2	df	Proportion
First dimension	(1) – (2)	92.28	10	75.97%
Second dimension	(2) – (5)	22.38	8	18.42%
Third and higher dimensions	(5)	6.81	10	5.61%
Total	(1)	121.47	28	

C. Collapsed table

	df	L^2	BIC	Δ	p
1. Independence model under (3)	9	105.10	38.59	8.87	.000
2. RC(1) model under (3)	4	12.87	–16.52	2.66	.012
3. Independence model under (6)	9	108.13	41.99	9.50	.000
4. RC(2) model under (6)	1	0.33	–7.02	0.22	.566

NOTE: See text for details.

123

Before we impose similar equality restrictions among row and/or column categories, we may want to compare the likelihood ratio test statistics of various models. Panel B provides the partitioned statistics for the analysis of association (ANOAS). While the first dimension captures about 76% of total association, the second dimension accounts for an additional 18.4%, leaving the remaining 5.6% to the third and higher dimensions. The partitioning result clearly indicates that the more complex formulation, $RC(2)$, should be preferred over its simpler counterpart, $RC(1)$. Note that the $RC(2)$ model has two cross-dimensional constraints $\left(\sum_{i=1}^{I} \mu_{i1}\mu_{i2} = \sum_{j=1}^{J} \nu_{j1}\nu_{j2} = 0 \right)$, in addition to centering and scaling restrictions, to uniquely identify all row and column score parameters.

By examining the converged estimates, Model 6 imposes the following equality constraints in both dimensions (row scores: ECO = MTO = MIL = ENR = SAB; column scores: ASAF = IFAA). Note that these constraints are not the same as those listed in Line 3. With a gain of 10 df, the model's deterioration of fit is marginal and insignificant ($L^2 = 6.8, p = .74$). In other words, the restricted $RC(2)$ model is clearly the preferred final model even though it does not have the most negative BIC statistic. Because the equality restrictions are imposed for both dimensions, the original 8×5 table can now be effectively reduced to form a new 4×4 aggregated table under $RC(2)$. This table has different cell frequencies when compared with another 4×4 aggregated table derived from the restricted $RC(1)$ model.

Panel C of Table 5.2 examines the consequences of adopting the wrong aggregated table in statistical modeling. Based on the new aggregated table from the restricted $RC(1)$ model, the independence model has 9 df, L^2 of 105. It appears that the difference in chi-square statistics between the current model and that of Model 1 under Panel A is rather small and acceptable (L^2 of 16.4 with 19 df). However, the $RC(1)$ model from Line C2 yields L^2 of 13 with 4 df and can be rejected at the 0.05 level. On the other hand, under the aggregated table derived from the restricted $RC(2)$ model, the results are very different. While the deterioration in goodness-of-fit between two independence models is also statistically insignificant (ΔL^2 of 13.4 with 19 df by comparing A1 with C3), the $RC(2)$ model in Line C4 provides satisfactory result as well.

A major merit of the restricted $RC(2)$ model is that it satisfies both the homogeneity and structural criteria simultaneously as described by Goodman (1981c) to determine whether certain categories can be combined. This model-based approach provides significant guidance to empirical researchers on how to minimize the loss in overall association while maintaining the underlying structure of association. In fact, previous studies have demonstrated that inappropriate or improper aggregation of categories can lead to distorted and erroneous understanding of the association between variables (Hou & Myles, 2008; Wong, 2003b). Unless there are good a priori reasons why certain categories should be combined, the best strategy is to use association models or other statistical models to understand if distortions have been introduced from improper aggregation and whether the latter would affect subsequent analyses.

The parameter estimates of the restricted $RC(1)$ and $RC(2)$ models are presented in Table 5.3. Because of the current limitation of the *gnm* module

Table 5.3 Selected Parameter Estimates of Principal Worries Example

		Parameter Estimates		
		(A) RC(1) Model With MIL = ECO = MTO, ENR = SAB = OTH, and ASAF = IFAA	(B) RC(2) Model With ECO = MTO = MIL = ENR = SAB, ASAF=IFAA in Both Dimensions	
Description		**First Dimension**	**First Dimension**	**Second Dimension**
ϕ_m		1.357	1.360	0.703
		(0.510)	(0.181)	(0.173)
μ_{im}	POL	−0.427	−0.549	−0.319
		(0.094)	(0.145)	(0.257)
	MIL	−0.173	−0.037	0.255
		(0.047)	(0.055)	(0.035)
	ECO	−0.173	−0.037	0.255
		(0.047)	(0.055)	(0.035)
	ENR	0.032	−0.037	0.255
		(0.046)	(0.055)	(0.035)
	SAB	0.032	−0.037	0.255
		(0.046)	(0.055)	(0.035)
	MTO	−0.173	−0.037	0.255
		(0.047)	(0.055)	(0.035)

(Continued)

126

Table 5.3 (Continued)

	Description	(A) RC(1) Model With MIL = ECO = MTO, ENR = SAB = OTH, and ASAF = IFAA	(B) RC(2) Model With ECO = MTO = MIL = ENR = SAB, ASAF=IFAA in Both Dimensions	
		First Dimension	**First Dimension**	**Second Dimension**
	PER	0.851	0.827	−0.236
		(0.040)	(0.117)	(0.254)
	OTH	0.032	−0.093	−0.720
		(0.046)	(0.148)	(0.145)
v_{jm}	EUAM	−0.699	−0.622	0.642
		(0.088)	(0.147)	(0.163)
	IFEA	−0.295	−0.400	−0.676
		(0.138)	(0.169)	(0.173)
	ASAF	0.457	0.473	0.159
		(0.068)	(0.099)	(0.137)
	IFAA	0.457	0.473	0.159
		(0.068)	(0.099)	(0.137)
	IFI	0.080	0.075	−0.284
		(0.223)	(0.237)	(0.294)

NOTE: Values in parentheses are the asymptotic standard errors under column (A) and bootstrap standard errors under column (B).

in R, asymptotic standard errors can only be obtained for the $RC(1)$ model. For the $RC(2)$ model, the bootstrap standard errors are reported instead. In addition, Figures 5.1 and 5.2 provide the graphical displays of estimated row and column scores from the unrestricted $RC(2)$ model (Line 5 of Table 5.2). These graphical displays will help decide which row and/or column categories are likely to have equal scores and therefore can be combined (Clogg & Shihadeh, 1994, p. 92). The key is to identify clusters or groupings of row categories or of column categories. From Figure 5.1, it is obvious that the following categories of principal worries have similar estimated row scores in the first dimension (MTO, MIL, SAB, ENR, ECO, and OTH). On the other hand, the estimated score for other worries (OTH) in the second dimension is much farther apart from the other five categories and therefore should not be combined together.

Similarly, the estimated column scores for living situations indicate that either IFI and IFAA or ASAF and IFAA forms a cluster of their own. To understand why the latter but not the former forms the correct cluster, one has to recall that the size of each $\hat{\phi}_m$ value will affect the variability observed along the different axes. "If $\hat{\phi}_m$ is large, there will be relatively

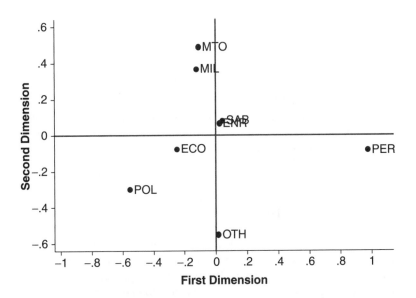

Figure 5.1 Estimated Scores for Principal Worries

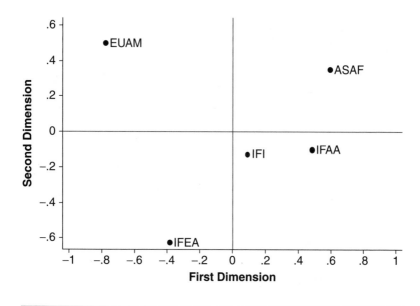

Figure 5.2   Estimated Scores for Living Places

large variability along the x-axis; if $\hat{\phi}_m$ is small, there will be relatively little variability along the y-axis or little vertical spread" (Clogg & Shihadeh, 1994, p. 99). Although the estimated row scores between IFI and IFAA in the second dimension are similar, they differ significantly in the first dimension. On the other hand, while the estimated row scores between IFAA and ASAF in the second dimension are quite dissimilar, its relatively small value of $\hat{\phi}_m$ means that the overall impact would be relatively small when the equality restrictions are imposed.

Example 5.2: Using Association Models as Optimal Scaling Tool

Although many social science applications naturally involve categorical variables, researchers often would like to treat or convert them into continuous measures so that they can be applied easily in other sophisticated multivariate models. This would pose an empirical dilemma to practitioners because we do not know what constitute the best assignment of scores to such variables. In this section, we make use of the "extrinsic" or "contingent" ordering properties of the estimated row, column, and layer scores from partial log-multiplicative association models to obtain the "desired" scores. It should be stressed that the choice of criterion variable(s) where such extrinsic orderings are obtained is crucial as the adoption of different criterion variables would yield different orderings. These chosen variables should be guided by sound theory and prior practices. The example below also extends existing literature by using more than one variable to yield more reliable and defensible scalings (Clogg & Shihadeh, 1994; Smith & Garnier, 1987).

The idea of using two or more criterion variables to obtain optimal scaling of another variable is not entirely new (note particularly the works of Duncan, 1982, 1984). In the case of socioeconomic status of occupations, for example, the Nam-Powers socioeconomic status scores are derived from the simple average of two separate rank orderings of incumbents' earnings and educational attainment from detailed occupational titles (Nam & Powers, 1983). A major problem of this calculation is that the derived measure is not model based and, therefore, does not necessarily maximize the relationship involved. The landmark work of Duncan (1961) adopts a different model-based approach that uses incumbents' education and earnings to predict occupational prestige and then interprets the predicted values as socioeconomic status scores. Despite its popularity, controversies remain as whether the resultant scores should be interpreted as status, prestige, or neither (Nakao, 1992). On the other hand, Clogg and Shihadeh (1994) hinted that it is possible to generate model-based measures within the framework of association models. However, their suggestion to scale individual occupation by using each criterion variable (education and earnings) separately and then taking averages between the two scores appears to be suboptimal (p. 58). While they also hinted that it is possible to obtain the desired scores simultaneously, albeit using slight different models, no further discussion has been provided. In the following example, we illustrate how this task can be achieved easily within the partial association framework outlined earlier.

Using the 1% 1980 Census data from the Integrated Public Use Microdata series (Ruggles et al., 2004), the second example cross-tabulates the major occupational groups for men and women aged between 20 and 64 with education and earnings (see Table 5.4). Occupation, as the row variable, has 12 major groups: (1) professional, (2) managerial, (3) sales, (4) clerical, (5) craftsmen, (6) operatives, (7) transport operators, (8) laborers, (9) farmers, (10) farm laborers, (11) service workers, and (12) private household workers. Education, as the column variable, has four categories: (1) less than high school, (2) high school, (3) some college, and (4) college or more. Finally, occupational earnings, as the layer variable, has four categories: (1) less than $1,000; (2) $1,000 to $5,999; (3) $6,000 to $9,999; and (4) $10,000 or more. For the sake of simplicity, the analyses below do not

Table 5.4 Major Occupation by Earnings by Education From 1970 Census (1% Sample)

Men and Women Aged 20–64	< High School				High School			
	<$1,000	$1,000 to $5,999	$6,000 to $9,999	$10,000+	<$1,000	$1,000 to $5,999	$6,000 to $9,999	$10,000+
Professional	1,096	1,847	1,255	925	3,321	6,123	6,830	5,524
Managerial	1,541	3,134	3,145	3,300	1,915	4,429	7,035	9,421
Sales	4,183	5,139	1,857	1,272	8,080	8,586	4,788	4,294
Clerical	6,033	9,211	5,046	1,058	28,130	44,589	20,074	3,408
Craftsmen	4,354	13,430	18,670	9,821	2,250	9,075	18,286	14,358
Operatives	14,587	31,470	16,390	3,751	8,242	17,532	12,825	3,956
Transport operators	1,517	5,820	6,197	2,372	721	2,909	4,141	2,070
Laborers	3,581	9,268	5,463	1,007	1,341	3,808	3,163	815
Farmers	1,454	3,109	1,055	888	563	1,909	1,018	1,051
Farm laborers	3,237	3,851	377	102	731	858	247	84
Service workers	14,882	22,182	5,363	1,136	11,650	15,818	5,524	2,122
Private household workers	6,033	3,475	63	18	1,603	1,005	30	16

	Some College				College+			
Professional	5,968	8,783	7,701	6,483	8,733	14,329	19,386	28,143
Managerial	1,011	2,162	3,536	6,649	697	1,299	2,362	10,796
Sales	3,214	3,621	2,485	3,177	793	1,134	1,292	3,597
Clerical	11,532	16,837	6,975	1,839	2,563	2,995	2,060	1,600
Craftsmen	1,009	2,719	3,521	3,409	296	503	626	1,273
Operatives	1,586	3,025	1,726	668	245	415	238	218
Transport operators	387	941	564	316	86	138	79	48
Laborers	994	1,988	542	145	158	259	101	56
Farmers	171	409	223	245	65	172	99	174
Farm laborers	293	290	67	31	32	62	18	30
Service workers	4,288	4,916	1,452	766	616	794	347	300
Private household workers	370	186	3	4	67	37	5	2

separate men from women and analyzes the total labor force to generate a single aggregated socioeconomic status index (TSEI) for the total labor force. The analyses below are therefore based on a three-way table ($12 \times 4 \times 4$), with a size of 819,798 individuals. It should be stressed that the analyses can easily be extended to include the more detailed occupational titles as well as four- or multiway classification tables, if needed.

Table 5.5 summarizes a series of statistical models applied to the three-way table. The first one is the complete independence model that yields extremely poor results (174 df, L^2 of 586,906), indicating some kinds of dependence among the three variables. The second model postulates conditional independence between education and income, once occupation is controlled. If the fit of the model is satisfactory, it would be consistent with the simple Markov causal chain as suggested in most stratification literature, that is, Education $\rightarrow$ Occupation $\rightarrow$ Earnings. Although the model (Line 2) offers a dramatic improvement, accounting for slightly more than 95% of the departure from complete independence, the relative fit of the model is still not satisfactory. The model with all two-way interaction between education, occupation, and earnings (Line 3), on the other hand, offers a better approximation of the relationship involved, accounting for an additional 3.6% of association (99 df, L^2 of 6,540).

The remaining models (Lines 4 to 8) attempt to decompose various partial association parameters into log-multiplicative components. For instance, Model 4 decomposes the occupation-education and occupation-earnings partial association, whereas Model 5 further imposes the restriction that the estimated scores for occupation in both partial associations are the same. According to conventional standards, the goodness-of-fit of both models is not satisfactory. Model 6 decomposes all three partial association terms into log-multiplicative components, whereas Model 7 imposes only equal row (occupational) scores, and Model 8 imposes equal row, column, and layer scores. Again, none of them provide satisfactory fit by conventional standards, but relative to others, Model 7 appears to be slightly preferred. Although it is possible to increase the dimensionality for all three partial association parameters, it is not performed here because our purpose is to extract maximal linear relationship between occupation, education, and earnings.

The estimated row (occupation) scores from Models 5, 7, and 8 are presented in Panel A of Table 5.6, and they are termed as TSEI1, TSEI2, and TSEI3, respectively. Since the estimated scores are standardized, they can be easily transformed into SEI-like measures by assigning specific values for their means and standard deviations. Consistent with our common understandings on occupational rankings, professional and managerial workers

Table 5.5 Association Models in Scaling Occupations Via Education and Earnings, 1970

Model Description	df	L^2	BIC	Δ	p
1. Complete independence	174	586906.22	584536.90	3.85	.000
2. Conditional independence	108	27957.40	26846.78	6.00	.000
3. All two-way interaction	99	6540.40	5192.33	2.64	.000
4. $RC(1) + RL(1)$ partial association	148	70860.99	68845.70	11.17	.000
5. $RC(1) + RL(1)$ partial association with consistent row (occupation) scores	158	185518.25	183363.80	18.27	.000
6. $RC(1) + RL(1) + CL(1)$ partial association	143	42101.44	40154.24	8.27	.000
7. $RC(1) + RL(1) + CL(1)$ partial association with consistent row (occupation) scores	153	174073.13	171989.76	17.80	.000
8. $RC(1) + RL(1) + CL(1)$ partial association with consistent row, column, and layer scores	157	177264.57	175126.73	17.76	.000

NOTE: The total sample size is 819,798.

133

Table 5.6 Selected Parameter Estimates of Association Models in Scaling Occupations

Description	Estimated Occupation Scores		
	RC(1)+RL(1)[a]	RC(1)+RL(1)+CL(1)[b]	RC(1)+RL(1)+CL(1)[c]
	TSEI1	TSEI2	TSEI3
A. Occupation scores			
Professional	0.548	0.569	0.575
Managerial	0.388	0.379	0.385
Sales	0.217	0.223	0.231
Clerical	0.144	0.171	0.175
Craftsmen	0.086	0.055	0.046
Operatives	−0.158	−0.164	−0.176
Transport operators	−0.050	−0.079	−0.097
Laborers	−0.141	−0.145	−0.160
Farmers	−0.024	−0.043	−0.045
Farm laborers	−0.399	−0.390	−0.388
Service workers	−0.105	−0.092	−0.090
Private household workers	−0.506	−0.484	−0.457

Estimated Occupation Scores

B. Correlations between TSEI scores, occupational prestige, and socioeconomic status

Description	RC(1)+RL(1)[a]		RC(1)+RL(1)+CL(1)[b]		RC(1)+RL(1)+CL(1)[c]	
	SEI	PRESTIGE	TSEI1	TSEI2	TSEI3	
SEI	1.000					
PRESTIGE	0.880	1.000				
TSEI1	0.925	0.913	1.000			
TSEI2	0.938	0.910	0.998	1.000		
TSEI3	0.946	0.913	0.995	0.999	1.000	

NOTE: The jackknife and bootstrap standard errors are not reported here because they are all very small (<.005). Also, R = Occupation, C = Education, and L = Earnings.

a. $RC(1)+RL(1)$ with consistent restrictions on row (occupation).

b. $RC(1)+RL(1)+CL(1)$ with consistent restrictions on row (occupation).

c. $RC(1)+RL(1)+CL(1)$ with consistent restrictions on row, column, and layer variables.

are clearly occupying the top of socioeconomic hierarchy whereas farm laborers and private household workers are at the bottom. Panel B presents intercorrelations between the three measures, together with the weighted Duncan's SEI and NORC's occupational prestige. With intercorrelations of 0.995 or higher, the three TSEI measures are virtually indistinguishable. That, of course, does not mean that there are no subtle differences. Although the ordering of the 12 major occupational groups are rather stable, the relative positions between transport operators and service workers can be slightly altered under different rankings.

It is interesting to observe that the correlation between SEI and prestige is rather low (0.88), a finding that is consistent with the assertion that "prestige scores are 'error-prone' estimates of the socioeconomic attributes of occupations" (Featherman & Hauser, 1976, p. 405). Viewed under this light, it is perhaps not surprising to find that the three derived TSEI measures have higher correlations with Duncan's SEI than occupational prestige. Their relatively high correlations with Duncan's SEI (greater than 0.92) suggest that the three derived TSEI scores share face and construct validity as a measure of socioeconomic status. In fact, everything being equal, the current measures should be preferred over SEI since the latter suffers from interpretational ambiguities as a prestige-derived measure of occupational status. The TSEI scores, on the other hand, are purely socioeconomic-based measure of occupation status. The proposed method is entirely consistent with the intents of the original proponents and users (see also the work of Ganzeboom, de Graaf, & Treiman, 1992; Ganzeboom & Treiman, 1996, for a slightly different construction method). The present technique has been applied successfully to create an indigenous SEI score of occupations in Hong Kong (Wong & Wu, 2006). In particular, the authors find that the constructed measures perform better than two other popular international measures: the standard international occupation prestige score (SIOPS) and international socioeconomic index (ISEI).

CHAPTER 6. CONCLUSIONS

The purpose of this monograph is to provide a systematic, coherent, and didactic introduction of different association models developed in the past two decades by a number of contributors, most notably the pioneering works of Leo A. Goodman and the late Clifford C. Clogg and Otis Dudley Duncan. Through a careful exposition of log-linear, log-multiplicative, hybrid, and multidimensional association models as well as their relationships among each other in two-way tables, the monograph then extends the analysis to three- and multiway cross-classification tables in subsequent chapters. When the three-way interaction terms are uninteresting and unnecessary, a series of statistical models that decompose some or all two-way interaction parameters are introduced. Otherwise, both two- and three-way (or higher-order) interaction terms can be decomposed under conditional association models. Through the use of numerous illustrative examples, readers should be able to gain a better appreciation of the power and flexibility of these association models.

It should be stressed that the $RC(M)$ and related association models for two- or higher-order cross-classification tables offer only one particular class of statistical models to understand relationships among a set of categorical variables in cross-tabular formats. Other (more or less) sophisticated statistical models may provide equally competing and sometimes perhaps even better understanding and should not be ignored in empirical investigations. This modeling strategy is consistent with the philosophy that it is only through the formulation of competing, alternative models that one can achieve insightful understanding of complex relationships among variables. Otherwise, one may risk choosing an incorrect model that may systematically distort the underlying relationship. On the other hand, if the goodness-of-fit statistics are satisfactory, one may find that the family of association models is particularly powerful and versatile to achieve enhanced understandings. In fact, the more complex the relationships are, particularly in three-way and multiway tables, the more likely that the multidimensional $RC(M)$-L association models with partial homogeneity or heterogeneity constraints can provide simpler and substantively interpretable results. The only issue is that we need to be careful about the need to impose all, some, or no cross-dimensional constraints and the proper accounting of the degrees of freedom of certain models.

The two practical applications further demonstrate that the RC-type association models can have wide applications in general social science research. In fact, similar RC-type association models have already been developed for latent structure analysis (Anderson, 2002; Anderson & Vermunt, 2000; Vermunt, 2001) and item response models (Anderson & Yu, 2007). Furthermore, it is well-known that the RC-type association models are closely related to correspondence analysis and canonical correlation analysis (Goodman, 1984, 1986). More recently, de Rooij (2008) and de Rooij and Heiser (2005) demonstrate that in the case of squared tables, when there is a one-to-one correspondence between row and column categories, the $RC(M)$ association model can be reparameterized as the two-mode distance association model and both models yield identical test statistics and expected frequencies.

Through the incorporation of the dynamic masses and dynamic positions assumptions to deal with asymmetric association that is common in social science applications, the generalized Newton's law of gravity model can be reparameterized as the $RC(M)$ association model (de Rooij, 2008). The parameters of μ_{im} and v_{jm} under Goodman's $RC(M)$ association model can be interpreted as within-set distances where an inner-product rule is used. In the case of the generalized gravity model, the corresponding distance parameters, z_{i1m} and z_{i2m}, can be interpreted as between-set distances. It is only the latter that provides the proper interpretation as distance measures. Furthermore, both sets of dynamic masses and dynamic positions parameters can be represented on a single graph to aid visual interpretation (de Rooij & Heiser, 2005).

Under specific circumstances, the generalized gravity model can be shown to be related to the partial association and conditional association models introduced here. The close relationship between natural science and social science formulations is certainly an intriguing development. Despite its allusion to natural science resemblance, the interpretation in terms of masses and distances is more as a *metaphor* than social reality. It is perhaps more fruitful to retain the original distinction of the association parameters as *patterns* and *levels* of association instead. Nonetheless, it is our hope that readers are now convinced and appreciative that association models can offer powerful and flexible ways to understand complex relationships among a set of categorical variables.

REFERENCES

Agresti, A. (1983). A survey of strategies for modeling cross-classifications having ordinal variables. *Journal of the American Statistical Association, 78,* 184–198.

Agresti, A. (1984). *The analysis of ordinal categorical data.* New York: Wiley.

Agresti, A. (2002). *Categorical data analysis.* New York: Wiley.

Agresti, A., & Chuang, C. (1986). Bayesian and maximum likelihood approaches to order restricted inference for models with ordinal categorical data. In R. Dykstra & T. Robertson (Eds.), *Advances in ordinal statistical inference* (pp. 6–27). Berlin, Germany: Springer-Verlag.

Agresti, A., Chuang, C., & Kezouh, A. (1987). Order-restricted score parameters in association models for contingency tables. *Journal of the American Statistical Association, 82,* 619–633.

Agresti, A., & Kezouh, A. (1983). Association models for multidimensional cross-classifications of ordinal variables. *Communication in Statistics, Series A, 12,* 1261–1276.

Aït-Sidi-Allal, M. L., Baccini, A., & Mondot, A. M. (2004). A new algorithm for estimating the parameters and their asymptotic covariance in correlation and association models. *Computational Statistics & Data Analysis, 45,* 389–421.

Andersen, E. B. (1980). *Discrete statistical models with social science applications.* Amsterdam: North-Holland.

Andersen, E. B. (1991). *The statistical analysis of categorical data.* Berlin, Germany: Springer-Verlag.

Anderson, C. J. (1996). The analysis of three-way contingency tables by three-mode association models. *Psychometrika, 61,* 465–483.

Anderson, C. J. (2002). Analysis of multivariate frequency data by graphical models and generalizations of the multidimensional row-column association model. *Psychological Methods, 7,* 446–467.

Anderson, C. J., & Vermunt, J. (2000). Log-multiplicative association models as latent variable models for nominal and/or ordinal data. *Sociological Methodology, 30,* 81–121.

Anderson, C. J., & Yu, J.-T. (2007). Log-multiplicative association models as item response models. *Psychometrika, 72,* 5–23.

Bartolucci, F., & Forcina, A. (2002). Extended RC association models allowing for order restrictions and marginal modeling. *Journal of the American Statistical Association, 97,* 1192–1199.

Becker, M. P. (1989a). Models for the analysis of association in multivariate contingency tables. *Journal of the American Statistical Association, 84,* 1014–1019.

Becker, M. P. (1989b). On the bivariate normal distribution and association models for ordinal categorical data. *Statistics & Probability Letters, 8,* 435–440.

Becker, M. P. (1990). Algorithm AS253: Maximum likelihood estimation of the $RC(M)$ association model. *Applied Statistics, 39,* 152–167.

Becker, M. P. (1992). Exploratory analysis of association models using loglinear models and singular value decompositions. *Computational Statistics & Data Analysis, 13,* 253–267.

140

Becker, M. P., & Clogg, C. C. (1989). Analysis of sets of two-way contingency tables using association models. *Journal of the American Statistical Association, 84*, 142–151.

Berkson, J. (1938). Some difficulties of the interpretation encountered in the application of the chi-square test. *Journal of the American Statistical Association, 33*, 526–542.

Bishop, Y. M. M., Fienberg, S. E., & Holland, P. W. (1975). *Discrete multivariate analysis: Theory and practice*. Cambridge: MIT Press.

Breen, R. (Ed.). (2004). *Social mobility in Europe*. London: Oxford University Press.

Carroll, J. D., & Chang, J. J. (1970). Analysis of individual differences in multidimensional scaling via an n-way generalizations of Eckart-Young decomposition. *Psychometrika, 35*, 283–319.

Choulakian, V. (1996). Generalized bilinear models. *Psychometrika, 61*, 271–283.

Clogg, C. C. (1982a). Some models for the analysis of association in multi-way cross-classifications having ordered categories. *Journal of the American Statistical Association, 77*, 803–815.

Clogg, C. C. (1982b). Using association models in sociological research: Some examples. *American Journal of Sociology, 88*, 114–134.

Clogg, C. C., & Rao, C. R. (1991). Comment on "Measures, models, and graphical displays in the analysis of cross-classified data." *Journal of the American Statistical Association, 86*, 1118–1120.

Clogg, C. C., Rubin, D. B., Schenker, D., Schultz, B., & Weidman, L. (1991). Multiple imputation of industry and occupation codes from Census Public-Use Samples using Bayesian logistic regression. *Journal of the American Statistical Association, 86*, 68–78.

Clogg, C. C., & Shihadeh, E. S. (1994). *Statistical models for ordinal variables*. Thousand Oaks, CA: Sage.

Clogg, C. C., Shockey, J. W., & Eliason, S. R. (1990). A generalized statistical framework for adjustment of rates. *Sociological Methods & Research, 19*, 156–195.

Davis, J. A., Smith, T. W., & Marsden, P. V. (2007). *General social surveys, 1972–2006* [Cumulative file] [Computer file]. ICPSR04697-v2. Chicago: National Opinion Research Center [producer], 2007. Storrs, CT: Roper Center for Public Opinion Research, University of Connecticut/Ann Arbor, MI: Inter-university Consortium for Political and Social Research [distributors], 2007-09-10.

de Rooij, M. (2008). The analysis of change, Newton's law of gravity and association models. *Journal of the Royal Statistical Society, Series A, 171*, 137–157.

de Rooij, M., & Heiser, W. J. (2005). Graphical representations and odds ratios in a distance-association model for the analysis of cross-classified data. *Pyschometrika, 70*, 99–122.

Diaconis, P., & Efron, B. (1985). Testing for independence in a two-way table: New interpretations of the chi-square statistic (with Discussion). *Annals of Statistics, 13*, 845–913.

Duncan, O. D. (1961). A socioeconomic index for all occupations. In A. Reiss Jr. (Ed.), *Occupations and social status* (pp. 109–138). New York: Free Press.

Duncan, O. D. (1979). How destination depends on origin in the occupational mobility table. *American Journal of Sociology, 84*, 793–803.

Duncan, O. D. (1982). *Rasch measurement and sociological theory*. Hollingshead Lecture, Yale University.

Duncan, O. D. (1984). *Notes on social measurement, historical and critical*. New York: Russell Sage Foundation.

Efron, B. (1981). Nonparametric estimates of standard error: The jackknife, the bootstrap, and other methods. *Biometrkia, 68*, 589–599.

Efron, B., & Tibshirani, R. (1993). *An introduction to the bootstrap*. New York: Chapman & Hall.

Eliason, S. R. (1990). *The categorical data analysis system, Version 3.50, User's manual* [Computer program]. Department of Sociology, University of Iowa.

141

Erikson, R., & Goldthorpe, J. H. (1992). *The constant flux: A study of class mobility in industrial societies.* London: Clarendon Press.

Featherman, D. L., & Hauser, R. M. (1976). Prestige or socioeconomic scales in the study of occupational achievements. *Sociological Methods & Research, 4,* 402–422.

Fienberg, S. S. (1980). *The analysis of cross-classified categorical data* (2nd ed.). Cambridge: MIT Press.

Firth, D., & de Menezes, R. X. (2004). Quasi-variances. *Biometrika, 91,* 65–80.

Fisher, R. A. (1925). *Statistical methods for research workers* (1st ed.). Edinburgh, UK: Oliver & Boyd.

Francis, B., Green, M., & Payne, C. (1993). *The GLIM system: Release 4 manual.* Oxford, UK: Clarendon Press.

Galindo-Garre, F., & Vermunt, J. K. (2004). The order-restricted association model: Two estimation algorithms and issues in testing. *Psychometrika, 69,* 641–654.

Ganzeboom, H. B. G., de Graaf, P., & Treiman, D. J. (1992). A standard international socioeconomic index of occupational status. *Social Science Research, 21,* 1–56.

Ganzeboom, H. B. G., & Treiman, D. J. (1996). Internationally comparable measures of occupational status for the 1988 international standard classification of occupations. *Social Science Research, 25,* 201–239.

Gilula, Z. (1986). Grouping and association in contingency tables: An exploratory canonical correlation approach. *Journal of the American Statistical Association, 81,* 773–779.

Gilula, Z., & Haberman, S. J. (1986). Canonical analysis of contingency tables by maximum likelihood. *Journal of American Statistical Association, 81,* 780–788.

Gilula, Z., & Haberman, S. J. (1988). The analysis of multivariate contingency tables by restricted canonical and restricted association models. *Journal of the American Statistical Association, 83,* 760–771.

Goodman, L. A. (1972). A general model for the analysis of surveys. *American Journal of Sociology, 77,* 1035–1086.

Goodman, L. A. (1974). Exploratory latent structure analysis using both identifiable and unidentifiable models. *Biometrika, 61,* 215–231.

Goodman, L. A. (1979a). Multiplicative models for the analysis of occupational mobility tables and other kinds of cross-classification tables. *American Journal of Sociology, 84,* 804–819.

Goodman, L. A. (1979b). Simple models for the analysis of association in cross-classifications having ordered categories. *Journal of the American Statistical Association, 74,* 537–552.

Goodman, L. A. (1981a). Association models and the bivariate normal for contingency tables with ordered categories. *Biometrika, 68,* 347–355.

Goodman, L. A. (1981b). Association models and canonical correlation in the analysis of cross-classifications having ordered categories. *Journal of the American Statistical Association, 75,* 320–334.

Goodman, L. A. (1981c). Criteria for determining whether certain categories in a cross-classification table should be combined, with special reference to occupational categories in an occupational mobility table. *American Journal of Sociology, 87,* 612–650.

Goodman, L. A. (1984). Some useful extensions of the usual correspondence analysis approach and the usual log-linear models approach in the analysis of contingency tables (with Discussions). *International Statistical Review, 54,* 243–270.

Goodman, L. A. (1985). The analysis of cross-classified data having ordered and/or unordered categories: Association models, correlation models, and asymmetry models for contingency tables with or without missing entries. *Annals of Statistics, 13,* 10–69.

Goodman, L. A. (1986). Some useful extensions of the usual correspondence analysis approach and the usual log-linear models approach in the analysis of contingency tables (with Discussion). *International Statistical Review, 54,* 243–270.

142

Goodman, L. A. (1987). New methods for analyzing the intrinsic character of qualitative variables using cross-classified data. *American Journal of Sociology, 93,* 529–583.

Goodman, L. A. (1991). Models, measures, and graphical displays in the analysis of contingency tables (with Discussions). *Journal of the American Statistical Association, 86,* 1085–1138.

Goodman, L. A. (2007). Statistical magic and/or statistical serendipity: An age of progress in the analysis of statistical data. *Annual Review of Sociology, 33,* 1–19.

Goodman, L. A., & Hout, M. (1998). Statistical methods and graphical displays for analyzing how the association between two qualitative variables differ among countries, among groups or over time: A modified regression-type approach. In A. E. Raftery (Ed.), *Sociological methodology 1998* (Vol. 28, pp. 175–230). Washington, DC: American Sociological Association.

Goodman, L. A., & Hout, M. (2001). Statistical methods and graphical displays for analyzing how the association between two qualitative variables differ among countries, among groups or over time. Part II: Some exploratory techniques, simple models, and simple examples. In M. P. Becker (Ed.), *Sociological methodology 2001* (Vol. 31, pp. 189–221). Washington, DC: American Sociological Association.

Greenacre, M. J. (1984). *Theory and applications of correspondence analysis.* New York: Academic Press.

Greenacre, M. J. (1988). Clustering the rows and columns of a contingency table. *Journal of Classification, 5,* 39–51.

Grusky, D. B., & Hauser, R. M. (1984). Comparative social mobility revisited: Models of convergence and divergence in sixteen countries. *American Sociological Review, 49,* 19–38.

Guttman, L. (1971). Measurement as structural theory. *Psychometrika, 36,* 329–347.

Haberman, S. J. (1978). *Analysis of qualitative data* (Vol. 1). New York: Academic Press.

Haberman, S. J. (1979). *Analysis of qualitative data* (Vol. 2). New York: Academic Press.

Haberman, S. J. (1981). Test of independence in two-way contingency tables based on canonical correlations and on linear-by-linear interaction. *Annals of Statistics, 9,* 1178–1186.

Haberman, S. J. (1995). Computation of maximum likelihood estimates in association models. *Journal of the American Statistical Association, 90,* 1438–1446.

Harshman, R. A. (1970). Foundations of the PARAFAC procedure: Models and conditions for an "exploratory" multi-modal factor analysis. *UCLA Working Papers in Phonetics, 16,* 1–84.

Harshman. R. A., & Lundy, M. E. (1984). Data preprocessing and the extended Parafac model. In H. G. Law, C. W. Synder Jr., J. A. Hattie, & R. P. McDonald (Eds.), *Research methods for multimode data analysis* (pp. 216–284). New York: Praeger.

Hauser, R. M. (1978). A structural model of the mobility table. *Social Forces, 56,* 919–953.

Henry, N. (1981). Jackknifing measures of association. *Sociological Methods & Research, 10,* 233–240.

Hou, F., & Myles, J. (2008). The changing role of education in the marriage market: Assortative marriage in Canada and the United States since the 1970s. *Canadian Journal of Sociology, 32,* 337–366.

Hout, M. (1983). *Mobility tables.* Beverly Hills, CA: Sage.

Hout, M. (1984). Status, autonomy, and training in occupational mobility. *American Journal of Sociology, 89,* 1379–1409.

Hout, M. (1988). More universalism, less structural mobility: The American occupational structure in the 1980s. *American Journal of Sociology, 93,* 1358–1400.

Ihaka, R., & Gentleman, R. (1996). R: A language for data analysis and graphics. *Journal of Computational and Graphical Statistics, 5,* 299–314.

143

Ishii-Kuntz, M. (1991). Association models in family research. *Journal of Marriage & the Family, 53,* 337–348.

Ishii-Kuntz, M. (1994). *Ordinal log-linear models.* Thousand Oaks, CA: Sage.

Kateri, M., Ahmad, R., & Papaioannou, T. (1998). New features in the class of association models. *Applied Stochastic Models Data Analysis, 14,* 125–136.

Kateri, M., & Iliopoulos, G. (2004). On collapsing categories in two-way contingency tables. *Statistics: A Journal of Theoretical and Applied Statistics, 37,* 443–455.

Knoke, D., & Burke, P. J. (1980). *Log-linear models.* Beverly Hills, CA: Sage.

Kotz, S., & Johnson, N. J. (Eds.). (1985). *Encyclopedia of statistical sciences* (Vol. 6). New York: Wiley.

Kruskal, J. B. (1977). Three-way arrays: Rank and uniqueness of trilinear decomposition, with application to arithmetic complexity and statistics. *Linear Algebra and Its Applications, 18,* 95–138.

Kruskal, J. B., Harshman, R. A., & Lundy, M. E. (1989). How 3-mfa data can cause degenerate PARAFAC solutions, among other relationships. In R. Coppi & S. Bolasco (Eds.), *Multiway data analysis* (pp. 115–130). Amsterdam: North-Holland.

Martin-Löf, P. (1974). The notion of redundancy and its use as a qualitative measure of the discrepancy between a statistical hypothesis and a set of observational data (with Discussion). *Scandinavian Journal of Statistics, 1,* 3–18.

Mooney, C. Z., & Duval, R. D. (1993). *Bootstrapping: A nonparametric approach to statistical inference.* Newbury Park, CA: Sage.

Nakao, K. (1992). Occupations and stratification: Issues of measurement. *Contemporary Sociology, 21,* 658–662.

Nam, C. B., & Powers, M. G. (1983). *The socioeconomic approach to status measurement: With a guide to occupational and socioeconomic status scores.* Houston, TX: Cap & Gown Press.

Pannekoek, J. (1985). Log-multiplicative models for multiway tables. *Sociological Methods & Research, 14,* 137–153.

Powers, D. A., & Xie, Y. 2000. *Statistical methods for categorical data analysis.* San Diego, CA: Academic Press.

Powers, D. A., & Xie, Y. 2008. *Statistical methods for categorical data analysis* (2nd ed.). Howard House, UK: Emerald.

Raftery, A. E. (1986). Choosing models for cross-classifications. *American Sociological Review, 51,* 145–146.

Raftery, A. E. (1996). Bayesian model selection in social research. In P. V. Marsden (Ed.). *Sociological methodology 1996* (Vol. 25, pp. 111–163). Washington, DC: American Sociological Association.

Raymo, J. M., & Xie, Y. (2000). Temporal and regional variation in the strength of educational homogamy. *American Sociological Review, 65,* 773–781.

Ritov, Y., & Gilula, G. (1991). The order-restricted RC model for ordered contingency tables: Estimation and testing for fit. *Annals of Statistics, 19,* 2090–2101.

Rosmalen, J. V., Koning, A. J., & Groenen, P. J. F. (2009). Optimal scaling of interaction effects in generalized linear models. *Multivariate Behavioral Research, 44,* 59–81.

Rudas, T. (1997). *Odds ratios in the analysis of contingency tables.* Thousand Oaks, CA: Sage.

Ruggles, S., Sobek, M., Alexander, T., Fitch, C. A., Goeken, R., Hall, P. K., et al. (2004). *Integrated public use microdata series: Version 3.0* [Machine-readable database]. Minneapolis: Minnesota Population Center [producer and distributor].

Siciliano, R., & Mooijaart, A. (1997). Three-factor association models for three-way contingency tables. *Computational Statistics & Data Analysis, 24,* 337–356.

144

Smith, H. L., & Garnier, M. A. (1987). Scaling via models for the analysis of association: Social background and educational careers in France. In C. C. Clogg (Ed.), *Sociological methodology 1987* (Vol. 17, pp. 205–246). Washington, DC: American Sociological Association.

Smits, J., Ultee, W., & Lammers, J. (1998). Educational homogamy in 65 countries: An explanation of differences in openness using country-level explanatory variables. *American Sociological Review, 63,* 264–285.

Smits, J., Ultee, W., & Lammers, J. (2000). More or less educational homogamy? A test of different versions of modernization theory using cross-temporal evidence for 60 countries. *American Sociological Review, 65,* 781–788.

Stegeman, A. (2007). Degeneracy in Candecomp/Parafac and Indscal explained for several three-sliced arrays with a two-valued typical rank. *Psychometrika, 72,* 601–619.

Tucker, L. R. (1966). Some mathematical notes on three-mode factor analysis. *Psychometrika, 31,* 279–311.

Turner, H. L., & Firth, D. (2007a). Generalized nonlinear models in R. *Statistical Computing & Graphics Newsletter, 18,* 11–16.

Turner, H L., & Firth, D. (2007b). gnm: A package for generalized nonlinear models. *R News, 7,* 8–12.

Vermunt, J. K. (1997). *LEM 1.0: A general program for the analysis of categorical data.* Tilburg University, Tilburg, The Netherlands. Retrieved September 1, 2009, from www.uvt .nl/faculteiten/fsw/organisatie/departementen/mto/software2.html

Vermunt, J. K. (2001). The use of restricted latent class models for defining and testing nonparametric and parametric IRT models. *Applied Psychological Measurement, 25,* 283–294.

Vermunt, J. K., & Magidson, J. (2005). *Latent GOLD 4.0 user's guide.* Belmont, MA: Statistical Innovations Inc.

Weakliem, D. L. (1992). Comparing non-nested models for contingency tables. In P. V. Marsden (Ed.), *Sociological methodology 1992* (Vol. 22, pp. 147–178). Oxford, UK: Basil Blackwell.

Weakliem, D. L. (1999). A critique of the Bayesian information criterion for model selection. *Sociological Methods & Research, 27,* 359–397.

Wong, R. S.-K. (1990). Understanding cross-national variation in occupational mobility. *American Sociological Review, 55,* 560–573.

Wong, R. S.-K. (1992). Vertical and nonvertical effects in class mobility: Cross-national variations. *American Sociological Review, 57,* 396–410.

Wong, R. S.-K. (1994). Model selection strategies and the use of association models to detect group differences. *Sociological Methods & Research, 22,* 460–491.

Wong, R. S.-K. (1995). Extensions in the use of log-multiplicative scaled association models in multiway contingency tables. *Sociological Methods & Research, 23,* 507–538.

Wong, R. S.-K. (2001). Multidimensional association models: A multilinear approach. *Sociological Methods & Research, 30,* 197–240.

Wong, R. S.-K. (2003a, March 1–3). *How sample size and strength of association affect the ability to detect group differences in cross-classification analysis.* Paper presented at the conference of the Research Committee on Social Stratification (RC28), International Sociological Association in Tokyo, Japan.

Wong, R. S.-K. (2003b). To see or not to see: Another look at research on temporal trends and cross-national differences in educational homogamy. *Taiwanese Journal of Sociology, 31,* 47–91.

Wong, R. S.-K., & Hauser, R. M. (1992). Trends in occupational mobility in Hungary under socialism. *Social Science Research, 21,* 419–444.

Wong, R. S.-K., & Wu, X. G. (2006, May 12–14). *Constructing an indigenous socioeconomic scale of occupation in Hong Kong: Issues and comparisons.* Paper presented at the International Sociological Association Research Committee on Social Stratification and Mobility at Nijmegen, The Netherlands.

Xie, Y. (1992). The log-multiplicative layer model for comparing mobility tables. *American Sociological Review, 57,* 380–395.

Xie, Y. (1998). Comment: The essential tension between parsimony and accuracy. In A. E. Raftery (Ed.), *Sociological methodology 1998* (Vol. 28, pp. 231–236). Washington, DC: American Sociological Association.

Xie, Y., & Pimentel, E. E. (1992). Age patterns of marital fertility: Revising the Coale-Trussell method. *Journal of the American Statistical Association, 87,* 977–984.

Yamaguchi, K. (1987). Models for comparing mobility tables: Toward parsimony and substance. *American Sociological Review, 52,* 482–494.

Yamaguchi, K. (1998). Comment: Some alternative ways to formulate regression-type models for three-way contingency table analysis to enhance the interpretability of results. In A. E. Raftery (Ed.), *Sociological methodology 1998* (Vol. 28, pp. 237–247). Washington, DC: American Sociological Association.

Yule, G. U. (1906). On a property which hold good for all groupings of a normal distribution of frequency for two variables, with applications to the study of contingency-tables for the inheritance of unmeasured qualities. *Proceedings of the Royal Society, Series A, 77,* 324–336.

Yule, G. U. (1912). On the methods of measuring association between two attributes. *Biometrika, 2,* 121–134.

AUTHOR INDEX

SUBJECT INDEX

Analysis of association (ANOAS):
  combining categories and, 124
  C+RC models and, 19
  RC models and, 17
Association parameters:
  education/occupation survey and,
    101, 104, 106
  intrinsic, 15, 18, 29, 57. *See also*
    Intrinsic association
    parameters
  Newton's unidimensional algorithm
    and, 16
Association patterns:
  association models and, 9, 119
  BIC statistics and, 27
  complex partial, 70, 74
  homogeneous association models
    and, 56
  isotropic, 78
  log-multiplicative row and column
    effects (RC) models and, 9
  model selection and, 26
  RC(2) models and, 20
  zero/sparse cells and, 30
Asymptotic standard errors, 39t, 49t
  cross-dimensional constraints and,
    52n 6
  jackknife standard errors and, 29
  Newton-Raphson algorithm and, 16,
    27–28
  reported standard errors and, 30
  software programs and, 25,
    52n 13

Bayesian Information Criterion (BIC):
  combining categories and, 121,
    122–123t, 124, 133t
  education/occupation survey and,
    100, 104
  education/premarital sex survey
    and, 109, 113t, 114
  one-dimensional association models
    and, 31, 33t
  saturated models and, 77
  selecting competing models and,
    26–27
  two-dimensional association models
    and, 42t, 43
Bayesian posteriori test theory, 26–27
Bootstrap standard errors, 29–30, 51n 1
  cross-dimensional constraints and,
    52n 2
  RC models and, 16, 17, 52n 13, 127
  See also Jackknife standard errors

CDAS, 16, 23, 24–25, 26
  web site for, 51n 4
  See also Software
Chi-square statistics:
  combining categories and,
    121, 124
  conditional independence
    parameters and, 63
  education/occupation survey and,
    101, 104
  education/premarital sex survey
    and, 109, 114

Note: In page references, f indicates figures and t indicates tables.

152

154

156

Supporting researchers for more than 40 years

Research methods have always been at the core of SAGE's publishing program. Founder Sara Miller McCune published SAGE's first methods book, *Public Policy Evaluation*, in 1970. Soon after, she launched the *Quantitative Applications in the Social Sciences* series—affectionately known as the "little green books."

Always at the forefront of developing and supporting new approaches in methods, SAGE published early groundbreaking texts and journals in the fields of qualitative methods and evaluation.

Today, more than 40 years and two million little green books later, SAGE continues to push the boundaries with a growing list of more than 1,200 research methods books, journals, and reference works across the social, behavioral, and health sciences. Its imprints—Pine Forge Press, home of innovative textbooks in sociology, and Corwin, publisher of PreK–12 resources for teachers and administrators—broaden SAGE's range of offerings in methods. SAGE further extended its impact in 2008 when it acquired CQ Press and its best-selling and highly respected political science research methods list.

From qualitative, quantitative, and mixed methods to evaluation, SAGE is the essential resource for academics and practitioners looking for the latest methods by leading scholars.

For more information, visit **www.sagepub.com**.